FOUNDATIONS OF INTERDISCIPLINARY AND TRANSDISCIPLINARY RESEARCH

A Reader

Edited by
Bianca Vienni–Baptista, Isabel Fletcher
and Catherine Lyall

With a Foreword by
Jane Ohlmeyer

BRISTOL
UNIVERSITY
PRESS

SHAPE-ID
Shaping interdisciplinary practices in Europe

First published in Great Britain in 2023 by

Bristol University Press
University of Bristol
1-9 Old Park Hill
Bristol
BS2 8BB
UK
t: +44 (0)117 374 6645
e: bup-info@bristol.ac.uk

Details of international sales and distribution partners are available at bristoluniversitypress.co.uk

Bristol University Press excluding Introductory Essay and Chapter 1 © Bianca Vienni-Baptista, Isabel Fletcher and Catherine Lyall, 2023

British Library Cataloguing in Publication Data
A catalogue record for this book is available from the British Library

ISBN 978-1-5292-2573-0 hardcover
ISBN 978-1-5292-2574-7 paperback
ISBN 978-1-5292-2575-4 ePub
ISBN 978-1-5292-3501-2 ePdf

Cover design: Andy Ward
Front cover image: iStock/brightstars

In memory of Julie Thompson Klein,
with gratitude for her friendship
and generous support.

Contents

List of Extracts

For a full list of copyright holders for the articles and chapters included in this book, please refer to 'Copyright Permissions' at the end of the book.

List of Acronyms

AAAS	American Association for the Advancement of Science
AHSS	Arts, humanities and social sciences
ECR	Early career researcher
ESRC	Economic and Social Research Council
HEI	Higher education institution
HEurope	Horizon Europe, European Union research and innovation funding programme
H2020	Horizon 2020, European Union research and innovation funding programme
ID	Interdisciplinarity
IDR	Interdisciplinary research
ITD	Inter- and transdisciplinary research
K★	Knowledge transfer, exchange, mobilisation
LERU	League of European Research Universities
MSCA	Marie Skłodowska-Curie Action
NGO	Non-governmental organisation
OECD	Organisation for Economic Co-operation and Development
OED	Oxford English Dictionary
PNAS	Proceedings of the National Academy of Sciences of the United States of America
RELU (or Relu)	Rural Economy and Land Use Programme
RMAs	Research managers and administrators
SHAPE-ID	Shaping Interdisciplinary Practices in Europe
SNSF	Swiss National Science Foundation
SSH	Social sciences and humanities
STEM	Science, technology, engineering and mathematics
STEMM	Science, technology, engineering, mathematics and medicine
STS	Science and technology studies
TD	Transdisciplinarity
td-net	Network for transdisciplinary research (Swiss Academies of Arts and Sciences)
TDR	Transdisciplinary research

Glossary

Agonistic   Used to describe conflicting attitudes. In this book it describes how researchers might establish a collaboration.

Artefact (artifact)   Objects created or built by researchers or artists. Artefacts are useful to learn about a group or a certain situation.

Bricolage   The construction or creation of a work from a diverse range of things (objects or ideas). In the humanities, the term is also used when groups borrow objects from others and create new aspects of their identities.

Consilience   Principle stating that several sources of evidence in agreement make evidence more robust. Reaching the same result applying different methods should lead to the same answer.

Constitutive   This term indicates an essential part of something, that is, a constituent.

Epistemology/epistemic   Epistemology refers to the theory of knowledge. It is concerned with questions such as: How do we know things? And if we do, how and when do we know things? Epistemic indicates the relation to knowledge.

Ethnocentrism   Mostly used in anthropology, an ethnocentric perspective is the evaluation of other cultures according to preconceptions originating in the standards and customs of one's own culture.

Ethnographic   Method used in anthropology to study other cultures by focusing on the scientific description of peoples and cultures with their customs, habits and mutual differences.

Formative (evaluation)   Assessment conducted during the development or improvement of a project or activity (in contrast to summative evaluation, which is conducted at the end of an activity).

Fungible   Something that can be substituted for something else.

Generative   Capable of producing or 'generating' something.

Heuristics   Guidelines that can be applied to aid decision making when information is limited.

Ideal-typical   Hypothetical mental construct representing a simplified version of reality, enabling comparison with real-life phenomena. An ideal-typical situation is neither 'perfect' nor an average, but an approximation to reality.

Meta–skills Short for 'metacognitive skills', higher-order skills that are applicable across domains and disciplines. An example could be communication skills.

Methodology (vs method) A method is a tool to answer research questions such as the technique used to collect data. A methodology is the rationale for the overall research approach, so it describes the overarching research strategy.

Normative When something (for example, a research finding) is compared with a (social) standard or 'norm'.

Ontology/ontological Branch of philosophy that analyses the nature of being and existence. In the social sciences, questions of ontology link to both epistemology and method since researchers' understandings of social reality affect the theoretical claims they can make.

Performative The concept that language (and by extension, other forms of behaviour) can function as a form of social action and thereby have effects on the world.

Positionality The social and political context that creates an individual's identity in terms of, for example, race, class, gender, sexuality and ability status. Also describes how that identity – derived from a social position – influences their understanding of and outlook on the world.

Positivist Positivism is an empiricist theory of knowledge, which holds that all genuine knowledge is true by definition, or derived by reason and logic from sensory experience ('positive').

Post–normal (science) Describes a problem-solving strategy appropriate in situations of urgency, uncertainty and disputed values, where standard processes of knowledge evaluation (such as risk assessment or cost-benefit analysis) fail. Climate change policy is an example of post–normal science.

Post–structuralism An intellectual movement that emerged in philosophy and the humanities in the 1960s and 1970s. It challenged previous ideas of structuralism, which believe that phenomena of human life are only understandable through their interrelations (such relationships constituting a 'structure').

Reflexive/reflexivity The capacity of an individual (often a researcher) to reflect on how their place in society has influenced their beliefs and behaviour, particularly when trying to make sense of their research data (see also 'Positionality').

Tacit Tacit knowledge is knowledge that has not been written down, codified or otherwise made explicit, making it difficult to communicate to others.

Wicked problem A problem that cannot readily be solved. There may be no single solution due to incomplete, contradictory and changing requirements, and the effort to solve one aspect of a wicked problem may reveal or create others.

Notes on Contributors

Maureen Burgess is Research Programme Officer in the Trinity Long Room Hub, Trinity College Dublin, where she supports arts and humanities researchers in securing competitive research funding. Maureen previously worked at the Higher Education Authority, the statutory policy-advisory body for higher education in Ireland, and the Irish Research Council for Humanities and Social Sciences. She holds a BComm with French from University College Cork and a Postgraduate Diploma in International Business Development from the Technological University of Dublin. She has an interest in exploring how arts and humanities disciplines can be meaningfully integrated in research addressing global challenges.

Kirsi Cheas is an Academy of Finland postdoctoral researcher at the Department of Communication, University of Vaasa, Finland and Visiting Scholar at New York University, Department of Media, Culture, and Communication, USA. She completed her PhD degree at the University of Helsinki in 2018. Her research focuses on collaborative investigative journalism, building on her broad experience in Latin American studies, media sociology and political communication. She is also founder and president of Finterdis, the Finnish Interdisciplinary Society.

Nathalie Dupin has a PhD in science and technology studies (STS) from the University of Edinburgh. She also holds a diploma in Systems Practice and a research Master's in STS. Her research interests centre on the higher education system, students' experiences, and the relationship between governmental policies and the production of academic knowledge. Her current research examines interdisciplinary doctoral training with a focus on changing academic practices. This study moves out of the lab to understand the nurturing of future scientists during their PhD programme, employing a practice theory perspective.

Isabel Fletcher is an interdisciplinary social scientist based at the University of Edinburgh's Global Academy of Agriculture and Food Systems and the Science, Technology and Innovation Studies subject group. Her research focuses on interactions between nutrition research and public policy, and the ways in which interdisciplinary and transdisciplinary research are used to address complex social problems. She was one of the co-creators of the SHAPE-ID (Shaping Interdisciplinary Practices in Europe) online toolkit of resources for inter- and transdisciplinary research, and is currently a co-investigator on the UK Research and Innovation-funded TRAnsforming the DEbate about Livestock Systems Transformation (TRADE) project.

Sabine Hoffmann is Group Leader for Interdisciplinary and Transdisciplinary Research at the Swiss Federal Institute of Aquatic Science and Technology (Eawag). She is also Head of the strategic inter- and transdisciplinary research programme on Water and sanitation innovations for non-grid solutions, WINGS. Her research focuses on inter- and transdisciplinary integration and integrative leadership. Her publications have appeared in *Science* (2016), *BioScience* (2022), *Environ Science & Technology* (2020), *Frontiers in Environmental Science* (2020), *Ecology & Society* (2019, 2017), *Environmental Science & Policy* (2021, 2019, 2018), *Humanities and Social Sciences Communications* (2022), *Issues in Interdisciplinary Studies* (2019), *Research Policy* (2017) and *GAIA* (2016, 2012). She was awarded with the Swiss Academies Award for Transdisciplinary Research.

Catherine Lyall is Professor of Science and Public Policy at the University of Edinburgh where, despite a first degree in chemistry, she holds a position in the School of Social and Political Science. She is an experienced science policy researcher and evaluator of knowledge exchange and interdisciplinary research activities who has acted as a consultant to public research funding organisations across the UK and Europe. She has brought this experience to bear in her previous publications, including *Being an Interdisciplinary Academic: How Institutions Shape University Careers* (Palgrave Pivot, 2019) and *Interdisciplinary Research Journeys* (with Ann Bruce, Joyce Tait and Laura Meagher, Bloomsbury Academic, 2011).

Jane Ohlmeyer, MRIA, FTCD, FRHistS, is Erasmus Smith's Professor of Modern History (1762) at Trinity College Dublin. She was the founding Head of the School of Histories and Humanities and Trinity's first Vice-President for Global Relations (2011–14). She was a driving force behind the 1641 Depositions Project and the development of the Trinity Long Room Hub Arts and Humanities Research Institute, which she directed (2015–20). She chaired the Irish Research Council (2015–21). She is the Principal Investigator for SHAPE-ID (Shaping Interdisciplinary Practices in Europe) and a Marie Skłodowska-Curie Action (MSCA) Co-fund called 'Human+', both funded by the European Commission's Horizon 2020 programme. She is the author and editor of numerous articles and 11 books.

She is currently working on a book on *Ireland, Empire and the Early Modern World*, which she gave as the Ford Lectures in Oxford (2021).

Christian Pohl is Adjunct Professor at ETH Zürich and Co-Director of the Transdisciplinarity Lab of the Department of Environmental Systems Science (www.tdlab.usys.ethz.ch). In his research, he studies transdisciplinary projects in the field of sustainable development, and explores tools to support knowledge co-production. His publications include the *Principles for Designing Transdisciplinary Research* (Oekom-Verlag, 2007) and the *Handbook of Transdisciplinary Research* (Springer, 2008). Together with td-net of the Swiss Academies of Arts and Science, he has developed a compilation of methods for co-producing knowledge, the td-net toolbox.

Sibylle Studer is Project Leader of the Focus 'Methods' (www. transdisciplinarity.ch/toolbox) at the Network for Transdisciplinary Research (td-net) of the Swiss Academies of Arts and Sciences. She is interested in toolkits, methods and capacity building as well as process design for co-production and related collaborative (research) modes. She is engaged in bridging interdisciplinary and transdisciplinary communities globally (for example, in https://itd-alliance.org). She has co-authored, with Christian Pohl, a forthcoming chapter entitled 'Toolkits for Transdisciplinary Research: State of the Art, Challenges, and Potentials for Further Developments' (in R.J. Lawrence [ed] *Handbook of Transdisciplinarity: Global Perspectives*, Edward Elgar Publishing Ltd).

Bianca Vienni-Baptista is Group Leader of 'Cultural Studies of Science' and Lecturer at the Transdisciplinarity Lab of the Department of Environmental Systems Science (USYS TdLab), ETH Zürich (Switzerland). Bianca works in the field of anthropology of science, focusing in particular on the study of collaborative knowledge production processes. As a result, she has centred her research on the specific conditions for interdisciplinary and transdisciplinary research and on the production and social use of knowledge in different countries, including the role of universities and other institutions. Her latest publications include *Institutionalizing Interdisciplinarity and Transdisciplinarity Collaboration across Cultures and Communities* co-edited with Julie Thompson Klein (Routledge, 2022).

Doireann Wallace is Senior Interdisciplinary Research Funding Specialist at Trinity College Dublin, where she supports interdisciplinary and transdisciplinary research development and funding applications in Climate and Sustainability areas, primarily for Horizon Europe. Doireann managed the SHAPE-ID (Shaping Interdisciplinary Practices in Europe) Horizon 2020 project at the Trinity Long Room Hub Arts and Humanities Research

Institute, contributing to developing the SHAPE-ID online toolkit and other publications. She has a PhD in media from the Dublin Institute of Technology, and has worked as an instructional designer, an art critic and a lecturer in photography and visual culture at the National College of Art and Design.

Acknowledgements

The editors would like to thank the following people, without whom this book would not have been possible:

- All of our SHAPE-ID partners, in particular Professor Jane Ohlmeyer, for bringing us together as a team, and for writing the Foreword.
- The members of the SHAPE-ID Expert Panel for their strategic input to the SHAPE-ID Toolkit (see www.shapeid.eu/expert-panel).
- Philippa Grand, for believing in the concept of a 'reader', and Paul Stevens and everyone at Bristol University Press for taking the project to fruition.
- The six peer reviewers who helped us to refine our initial ideas and provided encouraging feedback on the manuscript.
- Our commentary writers, for introducing broader perspectives and sharing their wisdom of practice.
- Nathalie Dupin, for input that went beyond editorial assistance.
- Dr Emily Woollen (Institute for Academic Development, University of Edinburgh), for her comments on the manuscript.
- The publishers involved in this project for their permission to reproduce the selected extracts (see 'Copyright Permissions' at the end of the book for full details).
- The editors of the 2015 collection *Encuentros sobre interdisciplina* (published by Espacio Interdisciplinario, Universidad de la República, Uruguay), who provided the inspiration for the current book.

The editors acknowledge support from the European Union's Horizon 2020 research and innovation programme under Grant Agreement No 822705.

Foreword:
Why Should We Care About Interdisciplinarity and Transdisciplinarity?

Jane Ohlmeyer

> The most exciting and ground-breaking innovations are happening at the intersection of disciplines. We need to cherish and encourage this as much as we can. But right now, our current infrastructure dissuades interdisciplinary research. (Carlos Moedas, EU Commissioner for Research, Science and Innovation, 2014–19)

As crises of democracy unfold around the globe, and as the climate emergency, the COVID-19 pandemic and the challenges associated with artificial intelligence (AI) all demonstrate, the problems of today are complex and multifaceted. Developing solutions demands a variety of disciplinary and practitioner approaches and perspectives. Interdisciplinary and transdisciplinary research provides radical solutions for some of these 'wicked' problems.

Shaping Interdisciplinary Practices in Europe (SHAPE-ID) was an EU-funded project that brought to the fore the importance of arts, humanities and social sciences (AHSS) perspectives as essential for addressing complex societal challenges in a more holistic and socially credible manner. It also offered very practical evidence-based recommendations about how we might better integrate the human and other sciences and collaborate. Thus, the project sought to improve pathways for interdisciplinary and transdisciplinary research (IDR and TDR), particularly between AHSS and science, technology, engineering and maths (STEM) disciplines.

This thoughtfully curated anthology builds on the learnings of the SHAPE-ID project. It brings together some of the most important literature by the global experts on interdisciplinarity and transdisciplinarity. There is a guided overview to each selection of readings, followed by a commentary from an expert or practitioner who ground the theory with their personal experiences of doing and supporting research that is inter- or transdisciplinary. Jargon-free and conversational in tone, this 'reader' is immediately accessible. It will prove an invaluable tool in the classroom as well as a 'how to' guide for those who want to 'dip their toes in the interdisciplinary pool' and to collaborate more effectively. It brilliantly complements and is complemented by the SHAPE-ID Toolkit, launched in June 2021, which provides a gateway to direct users to relevant guides, checklists, case studies and recommendations stemming from the project. It is also heartening to note how many of the contributors to this book are women. Female researchers and research officers, representing different domains and expertise, are empowering the conversations on inter- and transdisciplinarity.

As the contributions make clear, inter- and transdisciplinarity are not for the faint hearted. Despite the rhetoric, it is still an area that is poorly funded and has limited institutional support. In particular, the potential contributions of arts and humanities disciplines are poorly understood, something that SHAPE-ID addressed by launching the new toolkit, the first of its kind to focus on AHSS integration in Europe. It reflects the insights that SHAPE-ID has gathered from reviewing understandings and best practices through academic and policy literature reviews, a survey and interviews with European researchers, policymakers and societal partners, stakeholder workshops with 160+ expert participants, and Expert Panel consultation on all of the ways in which AHSS can contribute in the context of inter- and transdisciplinary research. Deep and meaningful collaboration with these areas is essential as researchers from different disciplines, with varied backgrounds and approaches, come together to frame problems jointly and to co-create research projects. It takes time to build trust, but it is remarkable how respect and reciprocity can prove transformative in creating ground-breaking possibilities: climate scientists working as equals with artists; philosophers with computer engineers; neuroscientists with dancers; and architects with non-governmental organisations (NGOs) – all to address some of the world's 'wicked' problems.

The nine chapters in this anthology mirror the nine major themes or 'pathways' in the SHAPE-ID Toolkit.[1] The challenges inherent in inter- and transdisciplinary research, by necessity, are to the fore, but so, too, are the solutions to overcoming them. These contributions, like the toolkit, bristle with constructive and practical suggestions about how best to navigate the slippery nomenclature that can often bedevil progress, and how to avoid instrumentalism and tokenism; how to create conditions that nurture respectful and inclusive collaboration and to build on the strengths that a multifaceted team enjoys; how best to evaluate inter- and transdisciplinary research and communicate the impact of it to other stakeholders, civil society, enterprise and policymakers.

The authors emphasise the importance of leadership; of creating research ecosystems that value disciplines alongside interdisciplinarity; of funding at an appropriate level the 'pipeline' of basic/curiosity-led and applied/challenge-led research; of supporting early career researchers, who are the agents of change and the life blood of our institutions; and of taking a long-term vision that creates a future research environment that is sustainable and thus better placed to serve the needs of society. The attributes, skills and expertise of individual researchers are also critical. Inter- and transdisciplinary researchers need to be open, curious and willing to listen and learn, to leave their egos at the door when meeting with others, and to be motivated to collaborate. Interdisciplinary teams also require a range of disciplinary expertise, integration expertise (people who can bridge the disciplinary knowledge) and also the involvement of relevant societal and enterprise stakeholders, as well as respect for their knowledge and experience.

Five key recommendations emerge from this book and the SHAPE-ID project where some of the thinking was formulated. These are:

1. *Promote socio-cultural missions and challenges:* Funders and policymakers need to commit to research and innovation missions driven by socio-cultural challenges and questions that foreground the human dimensions of challenges and put human flourishing at their centre.
2. *Co-design funding calls with AHSS experts:* AHSS researchers and societal stakeholders need to be consulted from an early stage in designing funding calls and programmes. Call language should be more open and inclusive, inviting a range of perspectives in addressing the topic and explicitly welcoming a broad range of contributions from AHSS and other stakeholders.
3. *Provide seed funding to enable relationships and capacity building:* Inter- and transdisciplinary research takes time and trust, and the AHSS disciplines in particular require support to build capacity. Seed funding is needed to consolidate networks, consortia and relationships, including industry and societal stakeholders, laying the foundations for larger-scale collaboration.
4. *Support a culture of inter- and transdisciplinarity in higher education:* Culture change takes time. Policymakers and funders can support the development of inter- and transdisciplinary education and research in higher education institutions to build capacity from undergraduate to postdoctoral and faculty level, training in 'meta-skills' and developmental support for those in institutional governance roles.
5. *Fund sustainable research careers, networks and infrastructures:* To facilitate knowledge sharing and community building across dispersed stakeholder groups, the European Commission, and other funding bodies, should provide sustainable funding for inter- and transdisciplinary infrastructure,

building on the SHAPE-ID Toolkit to create a more dynamic, interactive and sustainable resource. Continuity of funding support is also needed to enable researchers to build inter- and transdisciplinary careers, ranging from curiosity-led through to challenge-led research.

Let me conclude on a personal note. I have worked on these key recommendations and appreciate the value of this anthology to those embarking on a similar journey. In my capacity as Chair of the Irish Research Council (2015–21) we developed a range of programmes that promoted inter- and transdisciplinary research and nurtured collaborative research between AHSS and STEM.[2] In my capacity as Director of the Trinity Long Room Hub Institute of Arts and Humanities (2015–20), I worked with amazing researchers, especially early career scholars, from over 20 disciplines and with colleagues from other interdisciplinary institutes in neuroscience, computer science, environmental science and brain health.

Over the course of my career as a historian I have had the privilege of collaborating with researchers from across a wide variety of disciplines, especially computer scientists, and my research has benefited enormously from these interactions. Our collaboration, along with IBM and a small Irish company, began with the digitisation of an historical archive (the '1641 Depositions').[3] More recently, the focus has been the development of 'knowledge graphs' that allow for the interrogation of large bodies of historical data. This enables me to conduct meaningful research, and identify connections that are not otherwise apparent, or would require extensive time and research to reveal. Together, we also secured a Marie Skłodowska-Curie Action (MSCA) Co-fund grant, called Human+, which places the human experience at the centre of technological innovation.[4]

It was a particular honour to serve as the Principal Investigator for the SHAPE-ID project (2019–21) and to bring together a remarkable group of researchers and practitioners from a wide variety of disciplines and backgrounds.[5] It was inter- and transdisciplinarity in action. We all wanted to learn from best (and worst) practice, never to 'reinvent the wheel', to shape policy, and to make a difference for researchers and funders committed to inter- and transdisciplinary research. Of course, it took time to create common frameworks and understandings, to listen, and to co-create. Of real value was Christian Pohl's 'Theory of Change' and 'give and take matrix', where we detailed our expectations and developed mutual understandings at an early stage of our project. You can imagine, then, our collective delight when the person commissioned by the EU to review SHAPE-ID felt that the 'quality of the work produced is world class, positioning Europe as a leader in this area', and that the SHAPE-ID recommendations are 'very sensible, deserving careful consideration by policymakers'. This was a tribute to the intellectual curiosity of the diverse and hard-working team, to meaningful collaboration, and to the potential and value of inter- and transdisciplinarity, which this anthology demonstrates.

Notes

1. The SHAPE-ID Toolkit (www.shapeid.eu) builds on the expertise of many different partners: the development project was led by Professor Catherine Lyall and Dr Isabel Fletcher at the University of Edinburgh, with support from the SHAPE-ID team at Trinity College Dublin, ETH Zürich, the Institute of Literary Research of the Polish Academy of Sciences, ISINNOVA, and Dr Jack Spaapen.

2. For example, the New Foundations programme, which engaged over 200 community, voluntary and charity organisations, funded 356 projects and made an associated investment in excess of €7.5 million. See https://research.ie/funding/new-foundations (accessed 25/07/2022).

3. See https://1641.tcd.ie (accessed 26/07/2022).

4. Human+ has received funding from the European Union's Horizon 2020 research and innovation programme under the Marie Skłodowska-Curie Action (MSCA) Grant Agreement No 945447. See https://humanplus.ie (accessed 26/07/2022).

5. I would also like to take this opportunity to thank the editors for inviting me to contribute this Foreword. I am also deeply grateful, for their professionalism and friendship, to all members of the SHAPE-ID consortium – Bianca Vienni-Baptista, Anna Buchner, Isabel Fletcher, Giorgia Galvini, Catherine Lyall, Maciej Maryl, Christian Pohl, Jack Spaapen, Sibylle Studer, Carlo Sessa, Keisha Taylor-Wesselink and Piotr Wciślik; to our project managers, Caitriona Curtis and especially Doireann Wallace; and to colleagues at the Trinity Long Room Hub, especially Maureen Burgess, Niamh Brennan, Mary Doyle, Giovanna De Moura Rocha Lima, Eve Patten and Declan Whelan-Curtin.

.

Introductory Essay:
Shaping Interdisciplinary and Transdisciplinary Research

Bianca Vienni-Baptista, Isabel Fletcher and Catherine Lyall

Who are we?

At the start of 2019, in a time and place that feels quite different and far away now, the three of us came together to work as part of the SHAPE-ID (Shaping Interdisciplinary Practices in Europe) project. SHAPE-ID was a Coordination and Support Action funded by the European Commission under the Horizon 2020 Framework Programme Grant Agreement No 822705. The aim of the project was to review understandings and best practice of doing and supporting interdisciplinary and transdisciplinary research (IDR and TDR)[1] involving arts, humanities and social sciences (AHSS) disciplines alongside societal partners and researchers from the sciences, technology, engineering, mathematics and medicine (STEMM). The SHAPE-ID research consortium brought together interdisciplinary and transdisciplinary scholars and practitioners from six countries, several of whom have contributed to this book (see the commentaries in Chapters 3, 5 and 8).

The SHAPE-ID project's primary stakeholder groups were funders and policymakers, research performing organisations, researchers and research partners from enterprise or society. A core objective of the project was to deliver a toolkit and recommendations that would guide these decision makers and researchers, at different levels of the research and innovation system, towards successful pathways to integrating AHSS disciplines in IDR and TDR with STEMM disciplines, and societal partners, and thereby help key stakeholders to make better decisions and promote change in policymaking, funding and educational institutions.

The three of us worked closely on the production of the toolkit. Between us we represent different degrees of familiarity with the academic literature and with the research policy contexts across a range of countries and continents. We have researched, taught and led workshops on the topic of interdisciplinarity (ID) and transdisciplinarity (TD), and share a view that, if we are going to undertake IDR and TDR, then we want to build on good practice. We want to do this in a way that helps researchers surmount some of the typical entry barriers so that the field can progress more quickly and on more solid foundations. Our work with SHAPE-ID has shown that, within the rhetoric of research funding bodies and the strategic plans of research institutions worldwide, the same issues and challenges keep recurring. In large part this is because current academic literature on inter- and transdisciplinary research is dispersed across many different knowledge domains. The result is that scholarship is less cumulative and embedded than might be expected. This book aspires to rectify this situation by guiding readers through the basics.

Who is the book for and how could it be used?

Interdisciplinarity and transdisciplinarity are now common terms in research policy, and have a long history in a variety of research fields. These modes of knowledge production promise to solve complex and multidimensional problems and inform science policy at all levels of the European research system. Both are relevant to the European Framework Programmes that foster 'mission-oriented' collaborative research between academic and societal stakeholders in order to tackle global challenges. Nevertheless, interdisciplinary and transdisciplinary research are not new,[2] and are not solely embraced by researchers responding to external funding drivers: research has long had a tradition of 'borrowing' from other disciplines, and indeed, this is how disciplines develop and evolve (Klein, 1996). Yet, despite being currently in high demand, the practice of IDR and TDR is still not well understood, or at least it is understood in very uneven ways across different research communities.

We have designed this book as an introductory text for those new to the area of inter- and transdisciplinary research, and for those who have already dipped their toes into the interdisciplinary pool but are keen to learn more and perfect their craft. Our primary purpose in selecting and explaining the key texts that follow is to provide readers with an overview of the varied ways in which these approaches to research have already been developed and practised. We seek to provide new entrants with solid foundations on which to build their own IDR and TDR by sharing existing knowledge of how to successfully conduct collaborative forms of research and reduce the endless relearning or 'reinventing of wheels' – previously a feature of much that has been written on the topic.

What this book therefore seeks to do is to develop a better shared understanding of what IDR and TDR are (or could be), and to provide a 'grounding' in key works which might, in turn, help to speed up the development and dissemination of inter- and transdisciplinary knowledge. What the book offers is a carefully curated selection of key readings on inter- and transdisciplinary theory and practice, sharing current good practice and benchmarking the progression of thinking about IDR and TDR.

Research that spans different disciplines and sectors takes place in many different contexts, where similar activities may be given different labels. Key terms we are aware of include: collaborative research, interdisciplinary research, team science and transdisciplinary research. Add to this the general view that different disciplines are still at different stages in their understanding of the benefits and challenges of IDR and TDR (as we explain in more detail) and we have a very inconsistent and complex landscape for newcomers to navigate. This heterogeneity makes it especially hard to identify cross-cutting ideas and underpinning themes when conducting boundary-spanning research and making use of the results of such research. This book therefore introduces the reader to a large, unwieldy and diverse body of rich literatures (that overlap and indeed, sometimes, contradict) in what we hope are easily digestible, bite-size pieces. The reader is invited to follow the pathways offered by the chapters that follow and to search for more details in their own journey through the literature. Each chapter includes a 'References and further reading' section, giving readers who are curious the opportunity to explore similar resources and readings.

We envisage this book might be used in three different, but complementary, ways. The first would be as a teaching text in research methods courses for (post-)graduate students who are increasingly keen to undertake inter- or transdisciplinary research projects and need to be introduced to the specific skills required to undertake such research successfully. The second would be as body of practical knowledge to draw on at key points in the research process – for example when writing a research proposal – or as a resource to structure workshops on particular topics – perhaps the specific challenges of funding or evaluating collaborative research. Finally, we hope that individuals and groups will make use of it more informally at the beginning of a collaboration, either individually or in reading groups. However they use it – and there are certainly other ways that we have not thought of – we hope that all readers will employ the extracts, commentaries and further reading as a way of identifying the pieces of writing that are most relevant to their particular situation, enabling them to benefit, as we have, from the collective wisdom of those who have done this before.

In presenting this book as a 'reader' for those seeking to find out relatively quickly about the topic before embarking on their own deeper explorations, we are not proffering a polemic or position statement on what interdisciplinarity or transdisciplinarity could or should be. Nor are we attempting to create a new field. No academic arguments will be constructed (or demolished), and we are emphatically not trying to draw all the existing inter- and transdisciplinary scholarship into one mega 'synthesis'. Indeed, we have written elsewhere on the benefits of embracing this heterogeneity (see Vienni-Baptista et al, 2022). This is straightforwardly a guided introduction to a complicated and multifaceted area of research, recognising that there is no 'right' way to do it but offering lessons to newcomers based on what we have found helpful in our own development as interdisciplinary researchers. Above all, this is our personal selection of readings, spanning literature that has influenced and informed our own thinking as interdisciplinary and transdisciplinary practitioners at different stages in our careers as well as drawing on the further knowledge that we have gained as a result of working together with partners in the SHAPE-ID project.

How did we select the readings?

Inter- and transdisciplinary research pose several general challenges that are rooted in the nature of these phenomena. IDR and TDR are not established as well structured fields in the academic and policy literatures, and insights on them are scattered across unrelated bodies of literature. There are distinct pockets of expertise, for example in sustainability science or in the lab-based biomedical sciences, but knowledge derived from these fields may not be directly applicable to other quite different combinations of disciplines. Collaborations among and across different fields of knowledge imply dynamic intertwinements of concepts, data, methods, theories and experiments. Researchers and practitioners may contribute to co-producing knowledge in different ways, depending on their worldviews and interests in the problem.

Historically, STEMM disciplines have been dominant in IDR and TDR, and the SHAPE-ID project focused on better integrating the arts, humanities and social sciences within inter- and transdisciplinary research practice. Integrating knowledge from a wider range of different disciplinary contexts, which includes AHSS in a meaningful way, is an intricate task that requires careful selection, interpretation and translation for readers from different backgrounds and with different forms of research experience.

To address this intricate task, this book aims at connecting inter- and transdisciplinary research practice and theory to offer a supportive space for readers to learn. We have carefully selected readings that take into consideration the historical and geographical contexts of implementing and supporting collaborative research projects. The commentaries that accompany each chapter offer insight into how theory and practice play out through case study examples and personal experiences. Invited authors who contribute with each commentary also have different disciplinary backgrounds and expertise in working in inter- and transdisciplinary settings.

We based our selection of readings on an extensive literature review done by the SHAPE-ID team (for details of the review methods see Vienni-Baptista et al, 2019, 2020a). After building a robust sample of literature, the team aligned qualitative and quantitative methods to map understandings of IDR and TDR found in the literature. Datasets were created by querying scientific citation databases, supplemented by bibliographies prepared during a preliminary scoping analysis of inter- and transdisciplinary literature.

We compiled academic and policy literatures on inter- and transdisciplinarity, and critically examined these sources in order to: (1) map different approaches to the same topic across these two corpuses, and (2) bring together different theoretical perspectives (Burgers et al, 2019). Academic literature consists of peer-reviewed journal articles, book chapters and books on ID and TD, while policy literature included non-peer-reviewed documents contributing to debates in research policy.[3] In both these literatures concepts of ID and TD overlap significantly with normative accounts of how to conduct inter- and transdisciplinary research.

We scanned both bodies of literature to identify the different understandings of IDR and TDR and the factors contributing to their success or failure. We qualitatively analysed 121 scientific papers and 103 policy reports, including readings selected from a survey of members of the SHAPE-ID team together with a Delphi study performed annually by the Network for Transdisciplinary Research (td-net, Tour d'Horizon of Literature, Switzerland).[4] These datasets (containing over 5,000 items) were summarised and are accessible online in a Bibliography that is available from the SHAPE-ID Toolkit.[5]

During our study, we identified a set of challenges that characterise IDR and TDR that are shared in different geographical contexts. Such challenges include, among others, the lack of perceived legitimacy of inter- and transdisciplinary research as scientifically sound modes of knowledge production, the fragmentation of inter- and transdisciplinary communities of practice, differences in national and international policy and practice in their treatment and funding of IDR and TDR, a lack of status of AHSS disciplines in relation to STEMM contributions, and the need to defend AHSS' constitutive territory. We then explicitly worked towards addressing these challenges in the selected readings and also by encouraging invited authors to reflect on the constraints that they, themselves, might have overcome in order to pursue an inter- or transdisciplinary career or to support these types of collaborative research, as we explain in the following section.

What did we learn from SHAPE-ID and how is this reflected in the book?

Throughout our work in the SHAPE-ID project, we have argued that inter- and transdisciplinary research urgently need to be better supported in research, funding and policy institutions (Vienni-Baptista et al, 2020b; Fletcher et al, 2021). The paradox of interdisciplinarity (as Peter Weingart [2000] termed it 20 years ago) – whereby interdisciplinarity is often encouraged at the policy level but poorly rewarded – still challenges the establishment of cross-sectoral boundaries and connections. The role of AHSS disciplines in IDR and TDR raises particular questions about barriers to their integration.

From the different analyses the SHAPE-ID team conducted, three major insights (relevant for researchers, funders and policymakers alike) emerged:

1. An urgent need to acknowledge plural understandings of ID and TD and permit them to coexist in research (and funding) environments.
2. Recognition that the conditions that influence IDR and TDR are context-dependent: factors that hinder IDR and TDR can be transformed into enabling measures, even during the development of a research project.
3. A demand (and responsibility) to reassess AHSS roles and functions in IDR and TDR so that these disciplines can contribute fully in inter- and transdisciplinary settings.

In what follows, we explore these three insights and draw connections between the different analyses carried out during this phase of the SHAPE-ID project.

Acknowledgement and commitment to plural understandings of inter- and transdisciplinarity

The academic literature shows no agreement over the definitions of interdisciplinarity and transdisciplinarity. Rather, it shows plurality and overlapping conceptualisations, even contested and contrasting discourses when we take into consideration AHSS perspectives on ID and TD. Solving societal problems is seen as the main purpose of IDR and TDR, but other parallel discussions are taking place that provide alternative and substantial models of collaborative knowledge production processes. For instance, some AHSS communities are aligned to critical and philosophical discourses on ID and TD (see, for example, Extract 2.4 by James Leach and Extract 3.4 by Chris Rust).

We argue that, rather than develop new definitions, it is necessary to find connections between the diverse definitions of ID and TD that currently co-exist within the academic literature. The lack of connections between different communities results in a tendency to adopt a narrow approach whereby researchers ignore alternative collaborative pathways; this acts as an obstacle to further integration of AHSS disciplines in inter- and transdisciplinary research. Differences between academic fields with regard to methodologies and output modalities are obvious, but differences also exist between universities (some invest much more time, people and money in supporting IDR and TDR than others), and between countries (some have developed IDR and TDR policies at the national level, and some are less advanced in this area) (Spaapen et al, 2020).

Researchers and funders alike need to recognise that ID and TD are conducted for different purposes and are conceived in different ways, for example, as: (1) objects of study; (2) methods; and/or (3) phenomena that vary according to historical and geographical contexts (see Extract 1.2 by Andrew Barry et al).

We argue for a plural understanding of interdisciplinarity and transdisciplinarity because this could substantially improve inter- and transdisciplinary research policymaking and funding by giving institutions a greater appreciation of the conditions that are needed to support IDR and TDR in different contexts (Vienni-Baptista et al, 2022; see also Chapter 4 focusing on funding of IDR and TDR). In Chapter 1, we show these differences by means of a selection of extracts, a commentary by Isabel Fletcher and a list of 'References and further reading'. Allowing for the co-existence of plural understandings of ID and TD could also support early career researchers wanting to focus on IDR and TDR by making the pathways towards such a career more transparent (Lyall, 2019; see also Chapter 9 for a deep dive into this topic).

Acknowledging this urgency also implies commitments, responsibilities and specific actions from different societal actors and institutions. Actions to be implemented to promote a cultural change towards inter- and transdisciplinary research can include (based on Vienni-Baptista, 2023):

1. *Co-production of concepts:* to support the coexistence of different definitions that are context-dependent, researchers, funders and policymakers alike can develop co-production processes. Co-producing means simultaneous processes through which understandings of the world are built and related to representations, identities, discourses and institutions (Jasanoff, 2013). These can be developed during the research process or while elaborating funding schemes.
2. *Systematisation and traceability of a range of processes and practices:* to acknowledge that ID and TD imply different phenomena for different societal actors demands that all actors involved in IDR and TDR should develop processes to systematise these varied practices. This would

involve creating a 'memory' of IDR and TDR, including common understandings and agreements on what IDR and TDR are, the factors that hinder or help ID and TD development, how to better integrate AHSS and what methods and tools to use.

3. *Mapping of understandings:* to take into consideration that different modes of ID and TD exist, and these operate according to various logics. Mapping plural understandings, using different tools, leads to new spaces (epistemological, team-based, institutional, cross-sectoral) where IDR and TDR can be performed. In these spaces, AHSS disciplines can engage in new collaborative roles and functions.

Factors that affect inter- and transdisciplinary research are mutable

Factors that help successful inter- and transdisciplinary research as well as those that hinder such efforts are concrete realities. If we consider interdisciplinarity and transdisciplinarity as dynamic phenomena with multiple understandings and a heterogeneity of practices, trying to divide a list of factors into positive and negative conditions for research can be tricky.

In the academic literature, we identify 25 factors[6] influencing the outcomes of IDR and TDR. In the policy literature, four main factors are mentioned: (1) appropriate funding (see Chapter 4); (2) existing academic career structures (see Chapter 9); (3) the extended timescale required to conduct good quality IDR and TDR (see Chapter 2); and (4) recognition of key inter- and transdisciplinary skills (Lyall, 2019; see Chapter 7).

A promising finding on the factors that can help or hinder IDR and TDR collaboration is the indication from the literature that the same factor may act as either a barrier or an opportunity, depending on the circumstances within a project. This means that factors can be changed, transforming them from problematic to enabling during the research process. In part, this depends on what we value within research cultures and how far we are willing to go to change them: 'As a community we create our value systems. We can also alter them' (Lattuca, 2001: 264).

A demand (and responsibility) to reassess the roles and functions or the arts, humanities and social sciences in inter- and transdisciplinary research

'AHSS' is a problematic label, obscuring the differences between a set of disciplines with very different cultures and histories, and variations in methods, epistemologies and ontologies. Moreover, the model of inter- and transdisciplinary research as providing solutions for complex social problems – sometimes labelled 'mission-oriented research' – can be especially inhospitable to AHSS researchers due to its instrumental and technocratic approach to research (see Extract 4.3 by Julia Stamm).

The uneven representation of AHSS disciplines within inter- and transdisciplinary research projects needs to be recognised: while the SHAPE-ID findings confirm considerable levels of integration between disciplines from social sciences and environmental science, medicine and computer science, they also highlight the comparatively lower integration of arts and humanities disciplines with non-AHSS disciplines.[7]

Perhaps the biggest challenge for AHSS disciplines is to fight prejudice and misconceptions, among both researchers and policymakers (Spaapen et al, 2020). Findings from the SHAPE-ID project showed that the subordinate roles and functions assigned to AHSS disciplines discourage their greater involvement with STEMM disciplines in IDR and TDR. The problem has two aspects. On the one side, AHSS researchers have a responsibility to show more willingness to collaborate with other disciplines. On the other side, pro-active funders and policymakers also have a responsibility to change things for the better to support AHSS integration in IDR and TDR.

The academic literature and the case studies we have developed[8] also reveal a plethora of relationships between AHSS and other disciplines in inter- and transdisciplinary research. Transformative connections (which imply a change in disciplinary domains) and productive convergence (in which researchers integrate different types of knowledges), for instance, go beyond the instrumental function usually attributed to AHSS disciplines where, so often, they act in a subordinate role to the STEMM disciplines; for example, 'doing the public engagement' (Balmer et al, 2015; Fletcher and Lyall, 2021) once the 'real' research has been completed.

Significantly, even within the existing inter- and transdisciplinary literature, there are knowledge 'silos'. However, interdisciplinary research is still much more prevalent among different STEMM disciplines than between STEMM and the AHSS disciplines. Moreover, there is ample evidence that, when AHSS and STEMM disciplines do come together, the research agenda is (1) predominantly led by STEMM and (2) AHSS disciplines are chiefly represented by the social sciences (rather than the arts and humanities) and frequently by only a very limited sub-set of social science disciplines (such as economics) (Vienni-Baptista et al, 2020b).

AHSS disciplines have a relevant role to play and can contribute to consolidating a cultural change towards IDR and TDR development. What we seek to do here, with this book, is to redress this imbalance and provide a more comprehensive and diverse account of how IDR and TDR can flourish across all the disciplines. In the selected readings that follow, we have deliberately sought out extracts from journals and books that might be less accessible to, for example, natural scientists, and to introduce them in ways that render them more easily understandable. By providing additional lists of further reading for each chapter, we have tried to cover the same topic from a range of perspectives while acknowledging the inevitable imbalance resulting from the, as yet, uneven dissemination of inter-and transdisciplinary knowledge across different disciplinary traditions.

Contribution to state of the art

In the UK, where two of us are based, the research we do is increasingly influenced by 'the excellence turn' (Gläser and Laudel, 2016), whereby only exceptional research is deemed worthy of research funding. And this, in turn, is influenced by the UK's national quality assessment mechanism, the Research Excellence Framework or 'REF', which shapes the publications and other outputs and impacts we generate, based on that 'original' and 'excellent' research (Lyall, 2022). So, by its very nature, this book is unlikely to be considered 'REF-able' in UK academic parlance. While this 'Introductory Essay' draws substantially on findings from the SHAPE-ID project, the anthologised structure of the chapters that follow would not qualify as 'new' research. It is these strictures that render the REF highly problematic for certain disciplines and for interdisciplinary research in particular. Irrespective of these somewhat parochial concerns, we believe that this book exemplifies the general need to ascribe greater importance to the integration and application of existing knowledge (Frodeman, 2014) and not simply to cherish the traditional scholarship of new discoveries within a single discipline (Lattuca, 2001). In doing so, we believe this book fills an important need and provides a novel contribution to the state of the art in several respects.

First, it brings together, and therefore renders more accessible, a range of key texts from a variety of disparate literatures. One of the characteristics of the arts and humanities is that scholars are more likely to publish in book format, which, unlike publications from the natural and increasingly the social sciences, may be less readily available electronically. While one of the impacts of the COVID-19 pandemic has been that many wealthy university libraries have extended their electronic subscriptions, the same cannot be said worldwide. An old-fashioned 'reader' still has a role to play.

Not least, this carefully curated anthology format offers much more than a collection of readings that could be simply downloaded from journal repositories. It allows us to consider the differences between, for example, interdisciplinarity, transdisciplinarity and team science, and how they might relate to the different contexts in which they were developed. At face value, a novice may consider that these are all simply forms of collaborative research. Yet a more nuanced reading of the literature shows us that transdisciplinarity, for example, is much more than 'interdisciplinarity plus engagement with stakeholders' but is characterised by an ethos of openness to new encounters, co-production of knowledge and reflexivity (Pohl and Hirsch Hadorn, 2007). Unlike some forms of ID and team science, TD routinely makes use of a range of new research methods to address power dynamics within collaborative projects. The very notion of transdisciplinary research may be new to some readers, albeit that they may already have some familiarity with interdisciplinary research (see, for example, Lyall et al, 2015). The format of this book therefore enables us to highlight cross-cutting issues without minimising important differences between these approaches or trying to impose standardisation.

Developing the book out of our work on the SHAPE-ID Toolkit and the research that underpinned its development has also enabled us to identify gaps in current knowledge; for example, we were unable to find much existing information on budgeting for ID and TD. We have also reflected on how different country contexts matter (see also Vienni-Baptista and Klein, 2022), and how different disciplines approach similar challenges but feel the need to brand them with their own imprimatur. Again, this may be a feature of modern research cultures that, at the same time as we are recognising the benefits of cross-disciplinary collaboration, academia still feels the need to carve out its own specialisms. Current work on 'responsible research and innovation' (Felt, 2018) may be one such example, and surely has much to learn from existing scholarship on transdisciplinary research.

How is the book structured?

This book shares elements of its structure with the SHAPE-ID Toolkit, which was developed with a similar aim of providing a guide to existing resources on how to undertake inter- and transdisciplinary research.[9] The Toolkit was structured around a set of nine goals that followed the research process from the beginning to its end – from finding out about interdisciplinarity and transdisciplinarity to developing a career in the field – and we have used these goals as chapter headings for this book.

The design of the SHAPE-ID Toolkit was based on a synthesis of the findings of the literature reviews, survey, interviews, stakeholder workshops and Expert Panel consultation that constituted the main elements ('work packages', in EU funding terminology) of the research project. The Toolkit is structured in such a way as to act as a gateway to direct users to relevant information tailored to their specific interests and goals. It takes the form of 'guided pathways' that provide different points of access to enable customised user journeys based on the needs of individual Toolkit users. The web-based Toolkit was therefore designed with these features in mind, allowing different entry points and pathways depending on the profile of the user (for example, researcher, funder); their existing level of knowledge about inter- and transdisciplinary research; and the tasks they wish to accomplish (such as co-create a project, evaluate an IDR and TDR proposal).

Following this 'Introductory Essay', nine chapters introduce readers to selected core texts that underpin the current state of knowledge in the field. These mirror the guided task-based pathways that we created for the SHAPE-ID Toolkit. As such, they represent the core topics in ID and TD that cover the main challenges for newcomers but also for more experienced researchers and practitioners. Each chapter begins with a short introduction to the topic written by one of the editors, followed by three or four selected extracts (reproduced by permission of the original publishers, see 'Copyright Permissions' at the end of the book) that illustrate key learnings on the topic from the perspective of different writers. These readings are accompanied by short commentaries provided by invited contributing authors (and, in some cases, the editors) where the commentators reflect on why these readings are important to the practice of ID and TD. In doing so, they draw on examples of how they, themselves, have made use of specific approaches (conceptual or practical) described in the extracts, and offer insights into how useful or successful they were. Through these personal reflections, our contributing authors explain how their own experiences relate (or not) to the accounts given in our chosen extracts, offering the reader different standpoints and key takeaway points from each chapter. Every chapter also includes a short list of suggested further readings to offer additional perspectives.

In adopting this format, our goal is to offer an overview of the many and varied forms of ID and TD so that readers can build on these foundations in their own work. Rather than attempt to offer an overarching final conclusion, the book closes with a short 'Epilogue' authored by the three editors, with contributions from our commentary writers, where we reflect on our own experiences of inter- and transdisciplinary research and the process of producing this collection. Given the nature of this book, it is not our goal as editors to imply that this collection constitutes the only way to do these forms of research.

The commentaries as a conversation on inter- and transdisciplinarity

Inter- and transdisciplinary research are collective endeavours. Sharing the accumulated wisdom of those with practical experience of these kinds of research was of particular importance to us as an editorial team. We liked the idea of some form of dialogue that would test out our understandings of the significance of the readings we had chosen as emblematic of some of the issues that practitioners might encounter. This also enabled us to bring in a range of perspectives and, crucially, lived experiences. In this section, we offer a conversation that intertwines key insights from the chapters of this book. In such a conversation, we elaborate on cross-cutting topics throughout the chapters, showing that the book is rich in offering food for thought and open spaces for further explorations by the readers. Chapters can also be read separately – they all contain a narrative that guides the reader into the topic.

We, the editors, invited colleagues at different stages of their career and with different roles within academia to share their insights in the form of commentaries that accompany each of the chapters in this book. This group of contributing authors reflects diversity in their engagement and experience with inter- and/or transdisciplinary research. Authors hold a variety of leadership or mentorship roles, and strive for IDR and TDR as junior or senior researchers, project or programme managers in different countries. Their commentaries offer personal reflections, weaving a dialogue with the scientific literature embedded within the carefully selected excerpts contained in each chapter. These commentaries do not impose fixed formulas for successful collaboration; rather, they offer honest accounts of the challenges we face when we try to *define, co-create, design, evaluate, fund, communicate, improve, support and develop a career* in inter- and transdisciplinarity.

Chapter 1 focuses on disentangling some of the understandings of inter- and transdisciplinarity. Literature revolving around definitions of these terms has extensively discussed the differences in conceptualisations and explores how these definitions influence practices and research processes. The selection of extracts in this chapter provides the reader with a detailed and nuanced framework for understanding different models of collaborative research.

Closing Chapter 1, Isabel Fletcher offers an insightful text by exploring the extracts in the light of her own career. She positions herself as a researcher interested in 'interdisciplinary topics', and from that perspective, details a personal account of the academic literature on interdisciplinary research. From Fletcher's commentary, we learn what is common to a generation of researchers working in interdisciplinary settings: 'we arrive to the field because of our specific interests and motivation but not really knowing the corpus of knowledge available to guide our interdisciplinary practices'. We, interdisciplinarians and transdisciplinarians, then, are and were grateful for Julie Klein's books to light our path.

Another important observation Fletcher draws is her experience in the interdisciplinary field of science and technology studies (STS). She rightly points out that some fields or disciplines encourage methodological (and epistemological) pluralism. Should this be the rule for all disciplines faced with current societal challenges? The internal structure of disciplines influences the ways in which researchers working within them practise interdisciplinarity (and transdisciplinarity). At the end, Fletcher is right in signalling that the extracts in this chapter give us a vocabulary to better explain what we do when we do inter- or transdisciplinary research.

Chapter 2 turns our attention to how to develop collaborative conditions when working in inter- or transdisciplinary settings. The selected extracts build a strong connection with Chapter 1, when Isabel Fletcher enquires about the concept of integration in interdisciplinarity and transdisciplinarity, and how researchers need to take opportunities to collaborate and engage in more restricted ways, without undermining the potentials of inter- and transdisciplinary research. However, there is a rich literature that provides guidelines on how to improve our collaborative skills, competencies and mindsets. Once more, a plethora of terms is presented – integrating, co-producing, co-creating, interfacing; with nuances, they point to the primary role of listening, understanding and enjoying working with others.

Bianca Vienni-Baptista builds on experiences from EU researchers in the commentary that accompanies this chapter. From her study of the experiences of these researchers when building and being invited to participate in large consortiums, she and her colleagues identified a group of factors that either facilitated or hindered the potential of such collaborations. Constraints ranged from individual inability to share a common goal to institutional obstacles that do not account for the time frames required for inter- and transdisciplinary research. In order to overcome these difficulties, Vienni-Baptista argues that 'care' is an indispensable component of successful collaborations. Care, affection and emotion in interdisciplinary research constitute recent topics of inquiry, together with the relational aspects of identities in collaborative settings (see, for example, Smolka et al, 2020; Schikowitz, 2021). Care in collaboration extends the individual attitude of tenderness to a collective practice, in which outputs are desired but the process of collaborating already constitutes a result.

As our focus is also on arts and humanities integration in inter- and transdisciplinary research, Vienni-Baptista highlights some of the differences that need to be considered when researchers from these fields are participating in or leading collaborative processes. As mentioned before in this 'Introductory Essay', new forms of framing problems and integrating perspectives can be put into practice when the arts and humanities have a voice (a loud and clear one, but not a soft or instrumental one, as Fletcher indicates in Chapter 1).

Continuing the conversation, **Chapter 3** centres on how to co-create a research project. It brings to the fore an often-neglected aspect of inter- and transdisciplinary research: the power dynamics that are at play in collaborative settings. An increasing number of authors have consistently posed the question of power in research ('who is entitled to start a transdisciplinary process and why?'; see, for example, Schmidt and Neuburger, 2017). However, this still seems to be a domain of other disciplines (such as anthropology or science and technology studies). The arts also call attention to the issue of power and asymmetries. Meaning-making and meaning change in interventions and artistic work offers a good example of processes of collaboration – between the artist, the researcher and the communities – which questions who has the authority to understand a work of art.

Acknowledging and accepting the power differentials that exist in collaborative settings (see Extract 3.1 from Felicity Callard and Des Fitzgerald) might be a means to fight the 'utopia of co-creation', as Sibylle Studer indicates in her commentary in Chapter 3. She offers an informed perspective of how methods and tools used in inter- and transdisciplinary research projects have potentials but also may impose limits to what we can learn in such processes. Following the authors in the extracts, Studer strives for a combination of methods that allows researchers to think about when and in which phase they want to craft moments of co-creation, with whom they envisage interacting and with what intensity.

Studer's substantive experience in facilitating capacity-building workshops endows her with a rich perspective on how the question on methods can be deconstructed when the arts come into the conversation. Studer reflects on how methods and tools rely on their co-creators for continuous meaning-making. The need for attribution may put some co-creators more at risk than others, particularly if the former are deemed less powerful. Interestingly, Studer asks whether such uneven situations may lead to the decision not to co-create. Co-creation, as Studer expresses, is a multifaceted process, starting with agreeing on definitions, as shown in Chapter 1, and strengthening skills, as mentioned in Chapter 2 and discussed further in Chapter 7.

Chapter 4 moves a step further and explores funding collaborative research projects. Catherine Lyall reflects on how politics and power dynamics are still in place when funding agencies decide in what way and how to support inter- and transdisciplinary research. She observes that debates on interdisciplinarity and transdisciplinarity have not moved on, with extracts from later periods still discussing the same challenges we, researchers, face when gaining funding for our proposals. Lyall rightly observes that institutions around the world still do not develop systemic programmes of funding schemes for inter- and transdisciplinary research. Some examples give us hope, and efforts, such as the current Horizon Europe programme, serve as inspiration for future approaches.

But funding alone is not enough. Effective inter- and transdisciplinary programmes need integral support in the medium and long term. Lyall's commentary reminds us of the need for greater flexibility in inter- and transdisciplinary funding. Inter- and transdisciplinary research need spaces, symbolic and physical, to develop and change as they respond to the needs and perspectives of different disciplines and societal actors.

From this commentary, we are reminded that inter- and transdisciplinary processes need careful planning. Who initiates interdisciplinarity (in all its forms) is significant – whether it is initiated and driven mainly by science managers and policymakers or by researchers (see Extract 4.3 by Julia Stamm). This can influence research and its achievements. We can argue that this unsystematic implementation of funding schemes has been the norm for many years, and has led to the untidy and inconsistent ecosystem that we see around inter- and transdisciplinary research. What are funding agencies aiming for when funding inter- and transdisciplinary programmes without a consistent policy? We described a similar situation in the SHAPE-ID policy briefs in which we argued that the exclusion of contributions from the arts and humanities when addressing societal challenges means that inter- and transdisciplinary research do not achieve their full potential (Fletcher et al, 2021). One way out of this problem is to foreground the role of the integration expert in inter- and transdisciplinary research, as colleagues have recently argued for (Hoffmann et al, 2022).

The evaluation of inter- and transdisciplinary research projects is the topic of **Chapter 5**. Assessing collaborative settings, their outputs and impact, constitutes a challenging task. It implies responsibility from reviewers but also from institutions. In his commentary, Christian Pohl agrees with Katri Huutoniemi and Ismael Rafols (see Extract 5.2) that breadth, integration and transformation are three key aspects of inter- and transdisciplinary research and their quality.

We call the attention of the reader to the second element – integration – which is also a key element of evaluation processes. Pohl suggests that good inter- or transdisciplinary proposals should explicitly state what the applicants mean by integration (for example, reaching consensus, relating differing viewpoints), why this is the appropriate type of integration given the project's purpose, and how the applicants plan to achieve integration. Such processes can lead to the co-creation of new types of leadership (and of power relations?). In the future, assessment of leadership skills that support successful collaborative process should be part of the evaluation process in inter- and transdisciplinary research (see Extract 2.1 by Catherine Lyall et al). In her commentary for Chapter 4, Lyall asserted that other forms of collaboration and knowledge co-production are being ignored in evaluation processes. However, processes to assess these aspects of collaboration are currently in use. Examples include the Quality and Relevance in the Humanities (QRiH)[10] criteria or the Declaration on Research Assessment (DORA)[11] followed by the Swiss National Science Foundation when evaluating career paths.

Chapter 5 focuses mainly on the evaluation of research proposals, but we could also extend Pohl's concerns to peer review processes for scientific papers and other publications. Although some could argue that this is 'a whole different kettle of fish', both processes entail the operation of levels of authority and power that are not usually openly acknowledged within the academic system. In this book, **Chapter 6** focuses on communicating inter- and transdisciplinary findings, which indirectly relates to how these outputs are assessed.

Chapter 6 considers what constitute successful inter- or transdisciplinary publishing or knowledge transfer processes. While selecting the extracts that compose this chapter, we recognised that a more profound understanding of this process is still needed. There is little literature analysing the challenges of communicating co-created findings in inter- and transdisciplinary research. Although a seminal work by O'Rourke and co-authors was written in 2013, other relevant topics have been added to this discussion through the years. One example is the paper by Christian Pohl (Extract 6.2), which argues that transdisciplinary research can contribute to policymaking using a collective process involving multiple policy cultures.

Chapter 6 shows different levels and purposes of communication processes. Extracts address topics such as publishing inter- and transdisciplinary outputs, the potential transdisciplinarity has as a means of bridging communication with the policy sector, and the impacts of inter- and transdisciplinary research. In the accompanying commentary, Sabine Hoffmann acknowledges this eclectic set of topics in relation to the overarching aim of communicating science. Hoffmann adds new references that emphasise the need for more sustained and intense interactions – both formal and informal – between researchers and target groups to ensure greater use of inter- and transdisciplinary research findings.

Hoffmann rightly points out that interdisciplinarians and transdisciplinarians face the pervasive culture of 'publish or perish', reinforced by obstacles posed when seeking to publish in high-impact journals. Following Erin Leahey et al (Extract 6.3), Hoffmann shows that inter- and transdisciplinary research is a 'high-risk, high-reward endeavour' involving fewer published papers but higher visibility in the long run.

Chapter 7 highlights the skills that inter- and transdisciplinary researchers and practitioners develop in the course of their work. These encompass attributes such as leadership, communication, negotiation and integration, already mentioned in the previous chapters. In this case, the extracts allow the reader to better understand the cultural and emotional aspects of inter- and transdisciplinary research. As social activities, collaborative research practices are embedded in relational but also personal approaches to work. In authoring the commentary for this chapter, Nathalie Dupin draws on her experiences as an early career researcher. This standpoint confers real depth and value to Dupin's observations. From her perspective, new researchers may underestimate the skills involved in building multiple relationships in interdisciplinarity and transdisciplinarity. To explain this further, she structures her reflections around three concepts that most benefit newer researchers or those new to inter- and transdisciplinary research – learning, ethics and reflexivity.

In the last few years, an increasing number of inter- and transdisciplinary researchers in different scientific communities have come to address failures in their work (see Fam and O'Rourke, 2021). Dupin has the courage to also address delicate matters such as the necessity for 'ethical research within reflexive relationships', a topic that is still an 'elephant in the room' in inter- and transdisciplinary settings. She rightly indicates that remaining ethical in our relationships is a critical factor in building trust within a team.

Dupin concludes her piece by addressing the relevance of reflexivity in inter- and transdisciplinary research. We researchers usually take for granted the personal and collective processes of reflection we are embedded into during collaborative practices. In her commentary, Dupin discusses two interrelated levels: how individual group members need to learn from each other to carry out effective collaborative work, and the importance of an ethical stance in relationships to maintain safe spaces in which members can freely contribute their best efforts. These relationships are continually evolving practices that require us to be aware of, and monitor, our own actions in relation to others.

Chapter 8 continues this dialogue on interdisciplinarity and transdisciplinarity by centring the attention on how to support collaborative researchers. Researchers, practitioners, students and administrators are often faced with institutional encumbrances when trying to align structural norms to the daily practice of conducting inter- and transdisciplinary research.

Maureen Burgess and Doireann Wallace, colleagues from the SHAPE-ID consortium, played a substantive role in that project as financial manager and project manager respectively. They contribute the commentary to this chapter as important actors in inter- and transdisciplinary projects whose voices are usually unheard but are, in fact, indispensable to achieving a successful collaborative process. Burgess and Wallace build their reflections around two aspects: the need to understand, map and connect inter- and transdisciplinary expertise within and beyond higher education institutions, and the importance of long-term vision and leadership to build a culture supportive of inter- and transdisciplinary research.

Undoubtedly, institutions hold responsibility and power to enable more and better inter- and transdisciplinary research, as Christian Pohl indicates in his commentary in Chapter 5 or Catherine Lyall in hers in Chapter 4. But how to make institutions and their authorities listen to demands about recognising, nurturing and facilitating integration expertise? One way suggested by Burgess and Wallace is to acknowledge the difficult realities of undertaking and supporting inter- and transdisciplinary research, and to find ways to map and connect the tacit expertise that exists across the institution, among both researchers and professional staff. Burgess and Wallace offer recommendations at different levels of commitment but equally relevant to supporting interdisciplinarity and transdisciplinarity in institutional contexts, all of which can be summarised in the urgent need to consolidate a cultural change towards impactful collaborative research.

Completing the book, **Chapter 9** places the emphasis on future generations of researchers developing a career in inter- and transdisciplinary research. From mixed messages to lack of adequate supervision or mentorship, early career researchers shape the field of interdisciplinarity and transdisciplinarity while navigating dangerous waters. How do we better support them and encourage them to take risks in an academic ecosystem where more experienced researchers still find it hard to work collaboratively? The chapter collects four extracts from different authors working in varied contexts dealing with complementary obstacles to consolidating an inter- or transdisciplinary career.

The commentary in this chapter is authored by Kirsi Cheas, an early career researcher herself, who has taken on an active role in fostering empowering stories for junior researchers that go beyond obstacles and difficulties. She offers an honest account of the constraints she faced, balancing realism and reassurance in supporting the careers of other inter- and transdisciplinary scholars. Cheas acknowledges her own struggles, but confesses that those have helped her to develop the qualities of perseverance, resistance and tolerance for ambiguity and failure that she finds are fundamental for a career in inter- and transdisciplinary research. Should we argue that early career researchers need to experiment with inter- and transdisciplinary research but not pay the price of burnout, frustration and insecurity based on fixed-term contracts and uncertain career prospects? We read elsewhere that appreciative leadership (Whitney et al, 2010) is a means to strive for positive power – meaning that we can bring out the best in people and situations when we have the courage to inspire others. And this can be the key for early career researchers: offering them appreciative supervision that allows for their independence, while providing care and guidance.

Cheas is a clear example of such appreciation within inter- and transdisciplinary research. Her piece describes ways to build communities of practice but also demonstrates the need for commitment and bridges between existing networks. The positive examples she gives provide clear evidence that 'faith' also plays a role in science, and by believing in the capacities early career researchers bring to inter- and transdisciplinary research, we are supporting cultural change.

From the commentaries we learn that researchers and practitioners embrace heterogeneity in conducting, accompanying, supporting or promoting inter- and transdisciplinary research. In the chapters that follow, the reader will learn about inter- and transdisciplinary research in a different way: there are no fixed formulas, pre-digested definitions or ready-prepared recipes. The book offers highlights, contrasts, surprises and food for thought. We encourage readers to be inspired and, perhaps, to use this book to inspire others to embark on inter- and transdisciplinary research.

Technical note

We have described how we curated the collection of readings that follow. In each case we have acquired the rights to reproduce selected extracts from these published materials from the copyright holders (see 'Copyright Permissions' at the end of the book). We have endeavoured to reproduce these extracts faithfully from the originals. This means that we have not edited or corrected any errors in the selected text. This results in, for example, a mix of UK/US English spellings, some idiomatic use of language and occasional typographical errors that were in the originals. To avoid excessive copyright charges, we have generally decided against including any original figures in these extracts except where that was unavoidable to make sense of the reading. In some of the longer pieces such as book chapters, where we have selected non-contiguous text, this has been indicated using ellipses (see Extract 4.3, for example). Following some user testing, we took the decision as an editorial group not to include in this book any references, footnotes, endnotes or any cross-referencing from the extracts themselves. After some debate, we took the view that, including all of this material might risk overwhelming the reader and be counterproductive to the overarching goal of the anthology, which is to provide a manageable introduction to a potentially unwieldy topic. This does mean that, occasionally, readers will see a table or a figure cited that has not been included with the extract. However, full reference citations are given for all of the extracts that we have included, and we hope that readers will be motivated to seek out and read sources that have inspired them in full. For each chapter, we include a 'References and further reading' section that includes any additional references cited by the commentary writer together with a small selection of readings that the editors felt provided complementary perspectives from the longer SHAPE-ID bibliography. Finally, in addition to a 'List of Acronyms', we have included a short 'Glossary' of (predominantly) social science terms that appear in some of the extracts. This is by no means an exhaustive glossary, and nor are we claiming that these are the only interpretations available, but the definitions we offer may make this anthology more accessible to readers from other backgrounds.

Notes

[1] To aid readability we have tried to minimise the use of acronyms. Despite this, IDR and TDR and ID and TD are sometimes given as acronyms.

[2] The term 'interdisciplinary' was apparently first used in the 1920s (Klein, 1990).

[3] We did not include policy literature in this reader, but we provide relevant references in the 'References and further reading' section at the end of each chapter.

[4] https://transdisciplinarity.ch/en/publikationen/tour-dhorizon

[5] See www.shapeidtoolkit.eu/wp-content/uploads/2021/03/Guide-Annotated-Bibliography-Academic.pdf

6 For a detailed list of factors see Vienni-Baptista et al (2020b).
7 For a detailed explanation on the quantitative methods that we applied and the findings, please refer to Vienni-Baptista et al (2020b).
8 To illustrate the roles that AHSS research and creative practice can play in IDR and TDR, the SHAPE-ID Toolkit offers short accounts of innovative research projects and infrastructures. These case studies can be downloaded as PDF documents from: www.shapeidtoolkit.eu/case-studies
9 See www.shapeidtoolkit.eu
10 www.qrih.nl/en/about-qrih
11 https://sfdora.org

References and further reading

Balmer, A.S., Calvert, J., Marris, C., Molyneux-Hodgson, S., Frow, E., Kearnes, M., Bulpin, K., Schyfter, P., MacKenzie, A. and Martin, P. (2015) 'Taking roles in interdisciplinary collaborations: Reflections on working in post-ELSI spaces in the UK synthetic biology community', *Science and Technology Studies*, 28(3): 3–25, DOI:10.23987/sts.55340.

Burgers, C., Brugman, B.C. and Boeynaems, A. (2019) 'Systematic literature reviews: Four applications for interdisciplinary research', *Journal of Pragmatics*, 145: 102–9, https://doi.org/10.1016/j.pragma.2019.04.004

Callard, F. and Fitzgerald, D. (2015) *Rethinking Interdisciplinarity across the Social Sciences and Neurosciences*, London: Palgrave Macmillan UK.

Fam, D. and O'Rourke, M. (eds) (2021) *Interdisciplinary and Transdisciplinary Failures: Lessons Learned from Cautionary Tales*, Abingdon: Routledge.

Felt, U. (2018) 'Responsible Research and Innovation', in S. Gibbon, B. Prainsack, S. Hilgartner and J. Lamoreaux (eds) *Handbook for Genomics, Health and Society*, New York: Routledge, Chapter 14.

Fletcher, I. and Lyall, C. (2021) 'Stem cells and serendipity: Unburdening social scientists' feelings of failure', in D. Fam and M. O'Rourke (eds) *Interdisciplinary and Transdisciplinary Failures: Lessons Learned from Cautionary Tales*, Abingdon: Routledge, pp 45–61.

Fletcher, I., Lyall, C. and Wallace, D. (2021) 'Pathways to interdisciplinary and transdisciplinary research: The SHAPE-ID Toolkit', Policy Brief, Shaping Interdisciplinary Practices in Europe, https://zenodo.org/record/4922825#.Y8Fy6OLP1hE

Frodeman, R. (2014) *Sustainable Knowledge: A Theory of Interdisciplinarity*, New York: Palgrave Pivot.

Gläser, J. and Laudel, G. (2016) 'Governing science: How science policy shapes research content', *European Journal of Sociology*, 57(1): 117–68, https://doi.org/10.1017/S0003975616000047

Hoffmann, S., Deutsch, L., Klein, J.T. and O'Rourke, M. (2022) 'Integrate the integrators! A call for establishing academic careers for integration experts', *Humanities and Social Sciences Communications*, 9: 147, https://doi.org/10.1057/s41599-022-01138-z

Huutoniemi, K. and Rafols, I. (2017) 'Interdisciplinarity in Research Evaluation', in R. Frodeman, J.T. Klein and R.C.S. Pacheco (eds) *The Oxford Handbook of Interdisciplinarity* (Second edn), Oxford: Oxford University Press, Chapter 35.

Jasanoff, S. (2013) 'Fields and Fallows: A Political History of STS', in A. Barry and G. Born (eds) *Interdisciplinarity: Reconfigurations of the Social and Natural Sciences*, Abingdon: Routledge, pp 99–118.

Klein, J.T. (1990) *Interdisciplinarity: History, Theory, and Practice*, Detroit, MI: Wayne State University Press.

Klein, J.T. (1996) *Crossing Boundaries: Knowledge, Disciplinarities, and Interdisciplinarities*, Charlottesville, VA: University Press of Virginia.

Lattuca, L.R. (2001) *Creating Interdisciplinarity: Interdisciplinary Research and Teaching among College and University Faculty*, Nashville, TN: Vanderbilt University Press.

Leahey, E., Beckman, C.M. and Stanko, T.L. (2017) 'Prominent but less productive: The impact of interdisciplinarity on scientists' research', *Administrative Science Quarterly*, 62(1): 105–39, https://doi.org/10.1177/0001839216665364

Lyall, C. (2019) *Being an Interdisciplinary Academic: How Institutions Shape University Careers*, London: Palgrave Pivot.

Lyall, C. (2022) 'Excellence with Impact: Why UK Research Policy Discourages "Transdisciplinarity"', in B. Vienni-Baptista and J.T. Klein (eds) *Institutionalizing Interdisciplinarity and Transdisciplinarity: Collaboration across Cultures and Communities*, Abingdon: Routledge, Chapter 2.

Lyall, C., Meagher, L. and Bruce, A. (2015) 'A rose by any other name? Transdisciplinarity in the context of UK research policy', *Futures*, 65: 150–62, https://doi.org/10.1016/j.futures.2014.08.009

Pohl, C. (2008) 'From science to policy through transdisciplinary research', *Environmental Science & Policy*, 11(1): 46–53, https://doi.org/10.1016/j.envsci.2007.06.001

Pohl, C. and Hirsch Hadorn, G. (2007) *Principles for Designing Transdisciplinary Research*, Proposed by the Swiss Academies of Arts and Sciences, München: Oekom Verlag.

Schikowitz, A. (2021) 'Being a "Good Researcher" in Transdisciplinary Research: Choreographies of Identity Work Beyond Community', in K. Kastenhofer and S. Molyneux-Hodgson (eds) *Community and Identity in Contemporary Technosciences*, Sociology of the Sciences, Yearbook 31, pp 225–45, https://doi.org/10.1007/978-3-030-61728-8

Schmidt, L. and Neuburger, M. (2017) 'Trapped between privileges and precariousness: Tracing transdisciplinary research in a postcolonial setting', *Futures*, 93(October): 54–67, https://doi.org/10.1016/j.futures.2017.07.005

Smolka, M., Fisher, E. and Hausstein, A. (2020) 'From affect to action: Choices in attending to disconcertment in interdisciplinary collaborations', *Science, Technology, & Human Values*, 46(5), November, https://doi. org/10.1177/0162243920974088

Spaapen, J., Vienni-Baptista, B., Buchner, A. and Pohl, C. (2020) *Report on Survey among Interdisciplinary and Transdisciplinary Researchers and Post-Survey Interviews with Policy Stakeholders*, H2020 Project 'Shaping Interdisciplinary Practices in Europe', https://zenodo.org/record/3824727#.Y9-X_y8w3o8

Vienni-Baptista, B. (2023) 'Disentangling interdisciplinarity and transdisciplinarity: The beauty of differing definitions', in O. Pombo, K. Gärtner and J. Jesuíno (eds) *Theory and Practice in the Interdisciplinary Production and Reproduction of Scientific Knowledge: Interdisciplinarity in the XXI Century*, Cham: Springer, pp 59–74, https://doi.org/10.1007/978-3-031-20405-0_2

Vienni-Baptista, B. and Klein, J.T. (2022) *Institutionalizing Interdisciplinarity and Transdisciplinarity: Collaboration across Cultures and Communities*, Abingdon: Routledge.

Vienni-Baptista, B., Fletcher, I., Lyall, C. and Pohl, C. (2022) 'Embracing heterogeneity: Why plural understandings strengthen inter- and transdisciplinarity', *Science and Public Policy*, 49(6): 865–77, https://doi.org/10.1093/scipol/scac034

Vienni-Baptista, B., Lyall, C., Ohlmeyer, J., Spaapen, J., Wallace, D. and Pohl, C. (2020b) *Improving Pathways to Interdisciplinary and Transdisciplinary Research for the Arts, Humanities and Social Sciences: First Lessons from the SHAPE-ID Project*, Policy Brief, https://zenodo.org/record/3824954#.Y9-YNC8w3o9

Vienni-Baptista, B., Maryl, M., Wciślik, P., Fletcher, I., Buchner, A. and Pohl, C. (2020a) *Final Report on Understandings of Interdisciplinary and Transdisciplinary Research and Factors of Success or Failure*, https://zenodo.org/record/3824839#.Y9-YTS8w3o8

Vienni-Baptista, B., Maryl, M., Wciślik, P., Fletcher, I., Buchner, A., Wallace, D. and Pohl, C. (2019) *Preliminary Report of Literature Review on Understandings of Interdisciplinary and Transdisciplinary Research,* H2020 Project 'Shaping Interdisciplinary Practices in Europe', https://zenodo.org/record/3760417#.Y8AnWuLP1hE

Weingart, P. (2000) 'Interdisciplinarity: The Paradoxical Discourse', in P. Weingart and N. Sterh (eds) *Practising Interdisciplinarity*, Toronto: University of Toronto Press, pp 23–40.

Whitney, D., Trosten-Bloom, A. and Rader, K. (2010) *Appreciative Leadership: Focus on What Works to Drive Winning Performance and Build a Thriving Organization*, New York: McGraw Hill.

1

Understanding Interdisciplinary and Transdisciplinary Research

Chapter overview

Several different labels – inter-, multi- and transdisciplinary research, collaborative research and team science – are used to describe research across disciplines and sectors of society. These labels are often specific to particular contexts, for example transdisciplinary research is predominantly used within sustainability science and team science within medical research. This can create confusion and make it more difficult for those from outside these fields, such as arts, humanities and social sciences researchers and creative practitioners, to get a foothold in these projects.

Julie Thompson Klein was one of the key theorists of interdisciplinarity and transdisciplinarity.[1] In her chapter from the 2017 *Oxford Handbook of Interdisciplinarity* (Extract 1.1) she examines typologies of interdisciplinarity, identifying patterns of consensus and new developments. The importance of this piece lies in the way Klein identifies similarities and differences among multidisciplinarity, interdisciplinarity and transdisciplinarity. Klein's explanation of the historical development of ideas about interdisciplinarity and transdisciplinarity provides the reader with a detailed and nuanced framework for understanding different models of collaborative research.

In their influential article, Andrew Barry et al (Extract 1.2) analyse three interdisciplinary fields that span the boundaries between the natural sciences or engineering, on the one hand, and the social sciences or arts, on the other. The fields are: (1) environmental and climate change research; (2) ethnography in the IT industry; and (3) art–science collaborations. The authors reflect on interdisciplinarity from an innovative perspective, elaborating on three logics that influence collaborative research: the integrative-synthesis, subordination-service and agonistic-antagonistic modes of collaboration. The subordination-service mode, in particular, has been an influential model in critical accounts of collaborations between the natural and social sciences.

Our third set of extracts comes from an article by Philip Lowe et al (Extract 1.3), which reviews some of the key challenges for those trying to produce more impactful social science by engaging strategically with natural scientists. These authors argue that effective engagement depends on overcoming basic assumptions that have structured past collaborative interactions. The article is based on their participation in a major research programme that examined the different assumptions underlying knowledge claims in collaborations between social and natural scientists. As their main contribution, the authors draw out the lessons for social and natural science in cross-disciplinary engagements. Extract 1.3 examines the authors' account of the different motivations for undertaking interdisciplinary research.

Finally, Lisa Lau and Margaret Pasquini's article (Extract 1.4) uses data from a series of interviews with lecturers and students (mostly from the Department of Geography at Durham University) to discuss attempts to bridge gaps between the sciences and the social sciences, and between the social sciences and the arts. This material is a good example of the specific complexities that interdisciplinarity entails when integrating dissimilar disciplines. We have extracted a section describing interviewees' differing understandings of interdisciplinarity and of geography, which illustrates some of these complexities – for example, how a researcher sees their current discipline can have an important influence on their willingness to engage with other disciplines.

EXTRACT 1.1

Klein, J.T. (2017) 'Typologies of Interdisciplinarity: The Boundary Work of Definition', in R. Frodeman, J.T. Klein and R.C. Dos Santos Pacheco (eds) *The Oxford Handbook of Interdisciplinarity* (2nd edn), Oxford: Oxford University Press, Chapter 3.

3.2. Interdisciplinary Integration and Collaboration

The OECD definition of ID was wide, encompassing any interaction ranging from "simple communication of ideas to the mutual integration of organizing concepts, methodology, procedures, epistemology, terminology, data, and organization of research and education" (in Apostel 1972, p. 25). Simple communication, though, does not entail key traits that Burns and Lattuca argue constitute ID. Integrated designs prioritize focusing, blending, and linking. In education for instance, courses achieve a more holistic understanding of a cross-cutting question or problem by combining historical and legal perspectives on public education or biological and psychological aspects of human communication (Burns 1999, pp. 11-12; Lattuca 2001, pp. 81-83). Scope varies though, ranging from narrow to wide or broad ID depending on the number of disciplines involved and the compatability of their epistemological paradigms and methodologies.

Many believe that ID is synonymous with collaboration. It is not. However, heightened interest in teamwork to solve complex intellectual and social problems has amplified the connection while fostering greater attention to the interaction of cognitive and social integration. Degrees of cooperation differ, though. In Boden's concept of shared ID groups tackle aspects of a complex problem. Yet, collaboration does not necessarily occur. In contrast, cooperative ID requires teamwork, exemplified by the collaboration of physicists, chemists, engineers, and mathematicians in the Manhattan Project to build an atomic bomb and in research on public policy challenges such as energy and law and order (1999, pp. 17-19). Differences are further evident in methodological versus theoretical ID.

3.3. Bridge Building versus Restructuring

In 1975 the London-based Nuffield Foundation's Group for Research and Innovation identified two basic metaphors of ID – bridge building and restructuring. Bridge building occurs between complete and firm disciplines, while restructuring detaches parts of several disciplines to form a new coherent whole. A third possibility occurs when a new overarching concept or theory subsumes theories and concepts of several disciplines, akin to the notion of TD (Group for Research and Innovation, 1975, pp. 42-45). Landau, Proshansky, and Ittelson's typology of two phases in the history of interdisciplinary approaches in social sciences illustrates the difference between bridge building and restructuring. The first phase, dating from the close of World War I to 1930s, was embodied in the Social Science Research Council and University of Chicago school of social science. The interactionist framework at Chicago fostered integration, and members of the Chicago school were active in efforts to construct a unified philosophy of natural and social sciences. The impacts were widely felt, and occasionally disciplinary "spillage" led to formation of hybrid disciplines, such as social psychology and political sociology. However, traditional categories of knowledge and academic structures remained intact.

The second phase, dating from the close of World War II, was embodied in "integrated" social science courses, a growing tendency for interdisciplinary programs to become "integrated" departments, and the concept of behavioral science. Traditional categories anchoring disciplines were questioned and boundaries blurred, paving the way toward a new theoretical coherence and alternative divisions of labor. The behavioral science movement, in particular, sought an alternative method of organizing social inquiry rather than tacking imported methods and concepts onto traditional categories. In addition, the concept of "area" posited greater analytical power while stimulating a degree of theoretical convergence also potential in the concepts of role, status, exchange, information, communication, and decision-making (Landau et al. 1962, pp. 8, 12-17).

[...]

3.5. Transdisciplinarity

The recent ascendancy of TD is a prominent development in the history of ID. In the OECD typology, TD was defined as a common system of axioms that transcends the scope of disciplinary worldviews through an overarching synthesis, such as anthropology conceived as the science of humans. Three participants in the OECD seminar differed, though, in elaborating the concept. Jean Piaget treated TD as a higher stage in the epistemology of interdisciplinary relationships based on reciprocal assimilations. Andre Lichnerowicz promoted "the mathematic" as a universal interlanguage, and Erich Jantsch embued TD with social purpose in a hierarchical model of the system of science, education, and innovation (in Apostel 1972). Since then, the term has proliferated. Four major trendlines appear at present.

The first trendline is a contemporary version of the epistemological quest for systematic integration of knowledge. The quest for unity spans ancient Greek philosophy, the medieval Christian summa, the Enlightenment principle of universal reason, Hegelian philosophy, Transcendentalism, the search for unification theories in physics, and E. O. Wilson's theory of consilience. Reviewing the history of TD, Joseph Kockelmans (1979) found it has tended to center on educational and philosophical dimensions of sciences. The search for unity today, though, does not follow from a pregiven order. It must be continually "brought about," Kockelmans emphasized, through critical, philosophical, and supra-scientific reflection. It also accepts plurality and diversity, an underlying value of the Centre International de Recherches et Études Transdisciplinaire (CIRET). The center is a virtual meeting space for a new universality of thought and type of education informed by the worldview of complexity in science.

The second trendline is an extension of the OECD definition of synthetic paradigms. Miller defined TD as "articulated conceptual frameworks" that transcend the narrow scope of disciplinary worldviews. Leading examples include general systems, structuralism, poststructuralism, Marxism, phenomenology, feminist theory, and sustainability. Holistic in intent, these frameworks propose to reorganize the structure of knowledge by metaphorically encompassing parts of material fields that disciplines handle separately (1982, 21; see also Stribos, this volume). In the early twenty-first century a variant of this trendline emerged in North America in the concept of "transdisciplinary science" in broad areas such as cancer research. It is a collaborative form of "transcendent interdisciplinary research" that creates new methodological and theoretical frameworks for analyzing social, economic, political, environmental, and institutional factors in health and wellness (see Hall et al., this volume).

The third trendline is akin to critical ID. Transdisciplinarity is not just "transcendent" but also "transgressive." In the 1990s, TD began appearing more frequently as a label for knowledge formations shaped by critical imperatives in humanities, critiques of disciplinarity, and societal movements for change. Tracking the history of ID in Canadian Studies, Jill Vickers (1997) linked TD and "antidisciplinarity" with movements that reject disciplinarity in whole or in part, while raising questions of sociopolitical justice.

Examples include women's, native/aboriginal, cultural communications, regional, northern, urban, and environmental studies. Antidisciplinary positions have also moved beyond the academic sphere, favoring materials in ways dictated by students' own transdisciplinary theories, cultural traditions, lived experience, and connotations of "knowledge" and "evidence."

The fourth trendline prioritizes problem solving. It was evident in the late 1980s and early 1990s in Swiss and German contexts of environmental research. By the turn of the century case studies were reported on an international scale and in all fields of human interaction with natural systems and technical innovations as well as the development context. The core premise is that problems in the Lebenswelt – the lifeworld – need to frame research questions and practices, not disciplines. This connotation is strong in projects, such as Global TraPs (Global Transdisciplinary Processes on Sustainable Phosphorus Management), and in groups such as td-net (Network for Transdisciplinary Research). Co-production of knowledge with stakeholders in society is a cornerstone of this trendline, realized through mutual learning and a recursive approach to integration (see also Pohl et al., this volume).

The fourth trendline also intersects with two prominent concepts in the discourse of TD –"postnormal science" and "Mode 2 knowledge production." They stand in striking contrast to the intellectual climate of the 1970 OECD seminar, shaped by the organizing languages of logic, cybernetics, general systems theory, structuralism, and organization theory. Postnormal science is associated with TD because it breaks free of reductionist and mechanistic assumptions about how things are related and systems operate. "Unstructured" problems are driven by complex cause-effect relationships, and they exhibit a high divergence of values and factual knowledge. Hence, they are associated with the concept of "wicked problems" (see Bammer, this volume).

Gibbons et al. (1994) also proposed that a new mode of knowledge production has emerged. Mode 1 is characterized by hierarchical, homogeneous, and discipline-based work; Mode 2 by complexity, nonlinearity, heterogeneity, and TD. New configurations of research are being generated continuously, and a new social distribution of knowledge is occurring as a wider range of organizations and stakeholders contribute skills and expertise to problem solving. Gibbons et al. initially highlighted instrumental contexts of application, such as aircraft design, pharmaceutics, and electronics. Subsequently, though, Nowotny et al. (2001) extended Mode 2 theory to argue that contextualization of problems requires participation in the agora of public debate, incorporating the discourse of democracy. When lay perspective and alternative knowledges are recognized, a shift occurs from solely "reliable scientific knowledge" to inclusion of "socially robust knowledge."

EXTRACT 1.2

Barry, A., Born, G. and Weszkalnys, G. (2008) 'Logics of interdisciplinarity', *Economy and Society*, 37(1): 20–49.

Modes of interdisciplinarity

Much of the heat manifest in debates about interdisciplinarity stems from the potential for polarized judgements about the creative or repressive status of disciplinary knowledge. On one side are those for whom disciplines are generative and enabling, the repositories of a responsible kind of epistemological reflexivity. Marilyn Strathern gives voice to such a perspective when she writes that 'the value of a discipline is precisely in its ability to account for its conditions of existence and thus … how it arrives at its knowledge practices' (2004, p. 5). On the other side are those who see disciplines as 'inherently conventional', 'artificial "holding patterns" of inquiry' sustained by historical casts of mind 'that cannot imagine any alternatives to the current [disciplinary] regime'. In this view the significance of interdisciplinary research lies in the contrast with what are taken to be the more restrictive structures of disciplinary knowledge. Only interdisciplinarity holds out the promise of 'sustained epistemic change' (Fuller, 1993, n.d., pp. 1, 4).

In thinking about the relations between disciplinarity and interdisciplinarity, however, it would be a mistake to contrast the homogeneity and closure of disciplines with the heterogeneity and openness of interdisciplinarity. On the one hand, interdisciplinary research can involve hypostatization and closure, limiting as well as transforming the possibility for new forms, methods and sites of research (Weingart & Stehr, 2000; Strathern, forthcoming). On the other hand, disciplines themselves are often remarkably heterogeneous or internally divided (Galison, 1996b; Bensaude-Vincent & Stengers, 1996). Consider, for example, the differences between theoretical and experimental high-energy physics (Knorr Cetina, 1999) or between computational and laboratory medicinal chemistry (Barry, 2005). Even more radical internal differences exist between physical and human geography (Harrison et al., 2004) and between the sub-disciplines of anthropology (Lederman, 2005). Indeed, disciplines are routinely characterized by internal differences; the existence of a discipline does not always imply the acceptance of an agreed set of problems, objects, practices, theories or methods, or even of a shared language or common institutional structures.

Yet this heterogeneity is not necessarily a source of instability. In one account, 'the disunified, heterogeneous assemblage of the subcultures of science is precisely what structures its strength and coherence' (Galison, 1996a, p. 13). Disciplines exhibit clear inertial tendencies, and differences within them may exist over long periods of time. They may develop ways of translating across and negotiating internal boundaries; or chronic internal intellectual divisions may persist unaddressed through pragmatic working arrangements, or may even be collectively denied. Disciplines should not therefore be regarded as homogeneous, but as multiplicities or heterogeneous unities marked by differences which are themselves enacted in multiple ways (cf. Laclau & Mouffe, 1985,

p. 96). The existence of a discipline does imply a historically evolving and heterogeneous nexus of objects, problems, theories, texts, methods and institutions that are thought to be worth both contesting and defending. The boundaries of a discipline and the form in which it should exist, then, are in question and in play. Disciplinary boundaries and contents are neither entirely fixed nor fluid; rather, they are relational and in formation – dynamics captured by Stefan Collini in a powerful metaphor when discussing the emergence of cultural studies from its disciplinary progenitors: 'Cultural studies is part of the noise made by the great academic ice-floes of Literature, Sociology and Anthropology ... as their mass shifts and breaks apart' (1994, p. 3).

Further conceptual ground-clearing is necessary in the face of efforts to define three types of cross-disciplinary practice: interdisciplinarity, multidisciplinarity and transdisciplinarity. Commonly, a distinction is made between multidisciplinarity – in which several disciplines cooperate but remain unchanged, working with standard disciplinary framings – and interdisciplinarity – in which there is an attempt to integrate or synthesize perspectives from several disciplines. Ian Hacking, for instance, sets out the case for multidisciplinarity when he argues for 'collaborating disciplines that need not be interdisciplinary' and that presume a strong disciplinary base in the study of complex objects (Hacking, n.d.). Transdisciplinarity, in contrast, is taken to involve a transgression against or transcendence of disciplinary norms, whether in the pursuit of a fusion of disciplines, an approach oriented to complexity or real-world problem-solving, or one aimed at overcoming the distance between specialized and lay knowledges or between research and policy or 'decision-making in society' (Lawrence & Després, 2004, pp. 398-400). Transdisciplinarity is the term favoured by Nowotny et al. for the Mode-2 knowledge production characteristic of what they term a 'Knowledge Society': thus, '[i]ts reflexivity, eclecticism and contextualization mean that Mode-2 knowledge is inherently transgressive. ... [It] transcends disciplinary boundaries. It reaches beyond interdisciplinarity to transdisciplinarity' (Nowotny et al., 2001, p. 89). Whatever their descriptive uses, in general these definitional efforts have not proven generative in analytical terms. As Petts, Owens, & Bulkeley (in press, p. 8) note, the various definitions point to a spectrum: 'at its weakest, interdisciplinarity constitutes barely more than cooperation, while at its strongest, it lays the foundation for a more transformative recasting of disciplines.' We therefore take 'interdisciplinarity' as a generic term for this spectrum, while signalling salient issues from the definitional debate as they arise in our analysis.

How then can we conceptualize the relations between disciplinary and interdisciplinary forms of knowledge? Previous policy interventions and theoretical literatures on interdisciplinarity have tended to assume an integrative or synthesis model of interdisciplinarity, in which the interdisciplinary field is conceived in terms of the integration of two or more 'antecedent disciplines' in relatively symmetrical form (Tait & Lyall, 2001; Ramadier, 2004; National Academies, 2005, p. 26; Mansilla, 2006; Nowotny, n.d.). A major recent study of interdisciplinarity articulates this position clearly:

> In this integrative approach it is proposed that interdisciplinary work should be judged according to the criteria of the 'antecedent disciplines' and the value will be assessed

in terms of these additive criteria. ... In this study we defined 'interdisciplinary work' as work that integrates knowledge and modes of thinking from two or more disciplines. Such work embraces the goal of advancing understanding (eg explain phenomena, craft solutions, raise new questions) in ways that would have not been possible through single disciplinary means. (Mansilla & Gardner, n.d., p. 1)

This model has been performative. In climate change research, for example, it is thought that natural scientific and social scientific accounts of impacts might be integrated into a more general model, with social scientists providing an account of social factors ('society', 'the economy') which impact on climate change and are in turn impacted on by climate change (Jasanoff & Wynne, 1998, p. 3). The development of mathematical models provides one way in which such a synthesis can be achieved. It is worth noting, however, that, far from leading to the formation of new heterogeneous fields, the development of increasingly 'universal' models can lead to new kinds of closure effected through synthesis (Bowker, 1993). While the integrative mode can augur epistemic change, then, it does not guarantee it.

In our view, interdisciplinarity should not necessarily be understood additively as the sum of two or more disciplinary components or as achieved through a synthesis of different approaches. If we take the *integrative-synthesis mode* as a first type, we want to propose two additional ideal-typical modes of interdisciplinarity, both of which figure prominently in our research and which may coexist in some fields. In the second, *subordination-service mode*, one or more disciplines are organized in a relation of subordination or service to other component disciplines. This points to the hierarchical division of labour that characterizes many kinds of interdisciplinarity, an arrangement that may favour the stability and boundedness of component disciplines and inhibit epistemic change. In this mode the service discipline(s) is commonly understood to be making up for or filling in for an absence or lack in the other, (master) discipline(s). In some accounts the social sciences are understood precisely in these terms. They appear to make it possible for the natural sciences and engineering to engage with 'social factors' which had hitherto been excluded from analysis or consideration. Social scientists are expected to 'adopt the "correct" natural science definition of an environmental problem "and devise relevant solution strategies"' (Leroy, 1995, quoted in Owens, 2000, p. 1143, n. 3); or they may be called upon to assess and help to correct a lack of public understanding of science (Irwin & Wynne, 1996). One of the key justifications for funding art-science, particularly in the UK, has been the notion that the arts can provide a service to science, rendering it more popular or accessible to the lay public or publicizing and enhancing the aesthetic aspects of scientific imagery. Ironically, our research suggests that, in the microsocial space of interdisciplinary practice, the hierarchy entailed in the subordination-service mode can be inverted. In art-science, scientists sometimes adopt a service role for artist collaborators, providing resources and equipment to further a project conceived largely in artistic terms (cf. Born, 1995), while in the IT industry engineers may be called into the service of ethnographers.

In the third, *agonistic-antagonistic mode*, in contrast, interdisciplinary research is conceived neither as a synthesis nor in terms of a disciplinary division of labour, but as driven by an agonistic or antagonistic relation to existing forms of disciplinary knowledge and practice. Here, interdisciplinarity springs from a self-conscious dialogue with, criticism of or opposition to the intellectual, ethical or political limits of established disciplines or the status of academic research in general – a transposition on the plane of the politics of knowledge of Mouffe's (2005) stress on antagonism as constitutive of the political. This does not mean that what is produced can be reduced to these antagonisms. Through this mode we highlight how this kind of interdisciplinary field or practice commonly stems from a commitment or desire to contest or transcend the given epistemological and ontological assumptions of historical disciplines – a move that makes the new interdiscipline irreducible to its 'antecedent disciplines'. We will show, for example, how certain advocates of ethnography in the IT industry seek explicitly to constitute ethnography as a field which may be intellectually antagonistic both to existing sociological approaches to the study of technology (Randall, Harper, & Rouncefield, 2005) and to narrowly scientific and technical understandings of the properties and uses of technical objects and devices (Suchman, 1987; Nardi, 1996; Dourish, 2001).

Prominent in discussions of interdisciplinarity are two further methodological orientations which span the three modes. On the one hand, interdisciplinarity is commonly identified with problem-solving in response to new problems or objects that, it is believed, lie beyond the frame of existing disciplines. But rather than conceive of problems arising *de novo* and demanding interdisciplinary solutions, we should understand them as constituted as interdisciplinary problems relationally through dialogue or dissatisfaction with the problematics proffered by existing disciplines and institutions. The problem-focused, policy orientation of interdisciplinary environmental research, for instance, developed in conjunction with the constitution of multi-dimensional practical and political issues such as GMOs and climate change (Berkhout, Leach, & Scoones, 2005, p. 10). Some have argued additionally for the development of interactive methods involving government officials in research design and execution, thereby bringing research closer to the context of application in environmental policy-making (Turnpenny & O'Riordan, 2007, p. 103). On the other hand, rather than being object-oriented, interdisciplinarity can be practice-oriented in the sense that, where a disciplinary division of labour persists, cross-disciplinary collaboration is idealized as a value in itself, and one that outweighs any particular project (Born, 1995, chs 7, 8; Strathern, forthcoming). Commentaries on art-science, for example, sometimes portray the microsocial collaborative endeavour between artists and scientists as a crucible for creativity and as itself a focal value.

We have suggested that interdisciplinarity takes a range of forms with distinctive effects. While the discourse of Mode-2 alerts us to the importance of accountability in contemporary science policy, in its desire to discern a unitary epochal shift it collapses a number of alternative modes and trajectories of interdisciplinarity. The difference that environmental social science can make to natural-scientific environmental research, or that ethnography can make to computer-science-led design in industry or HCI (human-computer interaction) research, or that art-science collaborations can make to artistic or scientific practices cannot be understood solely in terms of making good an absence of connection to society, a lack of cognizance of users or a lack of public engagement with science. Rather, for some of their proponents such fields are intended to effect qualitative transformations, experimenting with and establishing new forms of practice that exist in an agonistic or antagonistic relation to, and that may destabilize, existing disciplines and practices. Yet while these kinds of interdisciplinarity cannot be cognized in terms of an additive synthesis of 'antecedent disciplines', and despite agonism or antagonism evident in a critique of disciplinary norms, a central concern of such research may well be strenuously to rebound on those antecedent disciplines, with the aim of reconfiguring their boundaries, objects and problematics.

If the integrative-synthesis mode can augur epistemic transformations, and if the service-subordination mode, with its disciplinary division of labour, does not necessarily afford even this, then what is striking about the agonistic-antagonistic mode is that it is intended to effect more radical shifts in knowledge practices, shifts that are at once epistemic and ontological. Indeed in what follows we propose that the three interdisciplinary fields that we studied evidence a privileged relation between the agonistic-antagonistic mode and the logic of ontology. To demonstrate this it is necessary to employ the framework outlined earlier, and specifically to do two things: first, through an account of the particular genealogies of each field, to indicate how the agonistic-antagonistic mode can only be understood diachronically in terms of a dynamic commitment to superseding prior ontological commitments with a new ontology; and, in doing so, to convey how this dynamic cannot be grasped by attributing a spurious unity. Instead, each interdisciplinary field must be analysed as precisely in play – as a heterogeneous unity or multiplicity.

EXTRACT 1.3

Lowe, P., Phillipson, J. and Wilkinson, K. (2013) 'Why social scientists should engage with natural scientists', *Contemporary Social Science: Journal of the Academy of Social Sciences*, 8(3): 207–22.

Motivations for interdisciplinarity

Among the Relu-funded ecologists, previous experience of interdisciplinary working varied from those who had an extensive history of collaboration with different types of social scientists to those for whom the Relu programme had provided a catalyst to work beyond their own field for the first time. The motivating factors cited by the ecologists map onto our three roles for social scientists.

Public representation was achieved by two mechanisms in the Relu projects: firstly, through the inclusion of social scientists, who necessarily provided a social dimension to the research through their understanding of social, political, regulatory and economic contexts, as well as through their data-gathering methods that allowed access to public views, opinions and knowledge. Additionally, each project was required to include a plan for stakeholder engagement, usually achieved through a set of advisors drawn from policy circles, community groups, the farming industry or other relevant audiences for the research. In practice, the two streams of public representation became blurred as researchers made creative use of their stakeholder networks through a variety of knowledge exchange activities and data gathering processes, which the social scientists were able to facilitate and analyse.

Several of the projects aimed to incorporate non-academic knowledge into their research, for example, by understanding how local communities perceive the risk of flooding (Lane et al., 2011) or how farmers interpret advice about farmland management and balance this against their own experiential knowledge (Proctor, Donaldson, Phillipson & Lowe, 2012). One of the ecologists described their motivation for working with social scientists:

> Social science plays a key part in our research because our project aims to combine knowledge from local stakeholders, policy-makers and social and natural scientists to anticipate, monitor and sustainably manage rural change in UK uplands. Key to this is linking the social and economic activities of local communities, through management, to the natural processes in upland landscapes. Without understanding these linkages policy prescriptions to influence management decisions may not have the anticipated ecological and social outcomes.

Another ecologist saw this desire to include stakeholder opinions as part of the broader trend of democratising science and breaking down the top-down model of knowledge transfer:

> The project is led by social scientists. The approach is to move away from black and white 'this is the science and this is what you need to do' towards involving the local community in deciding future actions based on good evidence.

The role of social scientists in problem framing became key as Relu funding bids developed, as researchers discovered the difficulty of designing projects from a monodisciplinary perspective and then trying to incorporate social science perspectives as an afterthought. As one ecologist commented:

> It is vital that both ecologists and social scientists have at least some understanding of how the other group thinks and works so some interaction before a project starts is necessary. Trying to respond to a call integrating social science and ecology without some prior interaction will probably result in failure to deliver. Understanding what each group requires of the other is also a key point to resolve at an early stage.

Joint problem framing was seen as critical to developing projects that would approach a key question or set of issues from multiple angles, ensuring a more coherent set of solutions could be delivered. To take one example, a project on organic agriculture aimed to understand the changing nature of agricultural production by jointly exploring both the socio-economic and the ecological factors driving, and being affected by, the uptake of organic farming. Two key questions were addressed: what causes organic farms to be arranged in clusters at local, regional and national scales, rather than be spread more evenly throughout the landscape; and how do the ecological, hydrological, socio-economic and cultural impacts of organic farming vary due to neighbourhood effects at a variety of scales. As a researcher on the project commented:

> [engaging with social scientists] places the natural science component in a context that will hopefully lead to meaningful policy decisions concerning sustainable agriculture and the multiple benefits that may accrue, only one of which is biodiversity. Without the social science perspective the natural science becomes rather meaningless.

Finally, researchers were motivated to engage in collaboration through a desire to more effectively understand and in some cases, impact upon the broader systems in which their research area was situated. Growing appreciation of the interrelationships between the social and natural dimensions of a problem led ecologists to seek the expertise of social scientists to maximise the utility of their research. In some cases, the expression of these aims came close to the end-of-pipe language of finding new ways to communicate science to non-experts, for example:

> The biological research is very applied with the aim to develop techniques/knowledge that can be applied. However, in the past uptake of such findings has often been poor. If we can better understand the constraints and forces driving farmers then we will be able to develop advice/techniques that fit within these.

However, a more nuanced approach emerged that recognised understanding interconnectedness as a way of doing science better, rather than simply having recommendations accepted more easily:

> It is all very well saying that a certain climate change scenario will lead to X, Y and Z biophysical consequences, but people live in that landscape and will adapt their behaviour to the changing climate in complex and dynamic ways. If we can capture this and understand how likely human responses will feed into the biophysical system, it is possible to provide a more nuanced, integrated and reliable assessment of future change.

These different comments reflect the continuing variation within the discipline of ecology with regards to the role that social science has to play. Within the survey as a whole, when asked how ecologists could more effectively address complex environmental problems, 44% felt that 'dealing more effectively with the social/human dimensions of their work' was what was primarily needed, while 35% felt they had to 'communicate their findings more effectively' and 22% thought the answer was to 'produce better ecological science' (see Lowe et al., 2009, p. 302).

For the social scientists, too, the contextual information provided by their natural science counterparts was invaluable in helping them to form a fuller picture of the problem they were investigating. Two political scientists commented (Greaves & Grant, 2010, pp. 332–333) that in both of the projects on biopesticides and livestock diseases they had been involved in

> the political scientists relied on the technical knowledge of the natural scientists to understand the precise nature of the policy challenges and the options open to the regulatory system to respond to them.

EXTRACT 1.4

Lau, L. and Pasquini, M.W. (2004) 'Meeting grounds: Perceiving and defining interdisciplinarity across the arts, social sciences and sciences', *Interdisciplinary Science Reviews*, 29(1): 49–64.

Our twin testimonies, describing interdisciplinary research spanning different academic spheres of knowledge, reveal that the sense of being an outsider is equally valid whether one is moving into the social sciences from the arts or from the sciences. The discovery of this commonality of experience prompted us to engage with notions of identity: our personal identities, our identities as geographers, our identities as interdisciplinary scholars, both as we ourselves perceived them and, importantly, as we deemed we were perceived by others. To this end, we carried out a series of interviews with fourteen respondents, who were chosen for their connections with and interest in interdisciplinary research. Seven of these were human geographers, five were physical geographers and two were anthropologists (selected for their close connections with geography). The geographers represented all five research groups within the Durham department – cultural and social geography, development studies (this group has since been dissolved), earth surface systems, political economies of geographical change and quaternary environmental change. The positionalities of the respondents greatly influenced their feedback, but in order to preserve their anonymity no further details can be revealed (including data on gender and position within the academic hierarchy). Instead, respondents have been given river names as pseudonyms.

The mention of interdisciplinary research generally brought about a deluge of positive comments. It was described as 'intellectually exciting', 'extremely interesting', 'stimulating intellectually, culturally in all sorts of different ways', 'fantastic, what we really need', 'of enormous value' and 'interesting'. Mississippi mentioned that over the last ten years there had been papers showing that the most productive work is done in marginal, interdisciplinary areas. Paraná said that the presence of interdisciplinary research made the department 'much more exciting, much more interesting, and of a much higher quality as a result of this, because the sparks fly, there is more electricity as a result of it, there's much more energy'. Interdisciplinary research encouraged people 'to challenge preconceptions' (Jamuna), it meant that 'research was better' (Jamuna) and it was definitely 'the way for the future' (Amazon).

The problem of understanding interdisciplinarity is akin to that of the proverbial onion: as we peel back layer after layer, so numerous complexities are revealed. Despite all the positive comments, because the interviews were designed to allow for individual, personal definitions and conceptions of interdisciplinary research, the respondents may have been talking at cross purposes and on different levels.

On the first level, there was contestation over the terminology. The geographers at Durham work with and from different definitions. Amu Darya felt that 'interdisciplinary' and 'cross-disciplinary' were synonymous, and that 'interdisciplinary' indicated links between two disciplines and 'multidisciplinary' between three or more. Respondents frequently commented that the approach to interdisciplinary research should be team based. Ganges felt that the word multidisciplinary should be used in preference to interdisciplinary because true interdisciplinarity can only be achieved if the partners in a research project work together side by side in the field for a long time. If monodisciplinary partners go into the field separately, and work on their own speciality, as normally occurs, the result is only multidisciplinarity. Missouri persisted in using the word interdisciplinary (which was interchanged with cross-disciplinary), but stressed that people had to work in a team, and that it was only after working in a team for many years that an individual could learn to see from alternative perspectives. Jamuna felt that a research programme can transcend disciplines, but that individuals tend to remain rooted in their specialisms, so 'multidisciplinary' is a more appropriate term. Rhine and Mississippi introduced the idea of 'post-disciplinary' research. A post-disciplinary world was understood to be topic driven (when disciplinary badges are set aside in order to work on a particular topic – such as feminism or post-colonialism – from different angles), and would be dominated by schools of thought rather than disciplines.

On a second level, there appeared to be roughly three camps of thought regarding geography and interdisciplinarity: those who were thinking in terms of links between geography and other disciplines, those who were thinking in terms of links within geography between different geographical research groups, and those who reflected on both. Implicit or explicit definitions of what constitutes interdisciplinary research quite naturally influenced the tenor of interview discussions. This is arguably of some importance: because there is no clear definition or consensual understanding of what constitutes interdisciplinary research, it is therefore not easy for geography to promote or support such work.

If interdisciplinary research is regarded as geography linking with other disciplines, then most of geography should be considered interdisciplinary. A large proportion of physical geographers are biologists, geologists, geophysicists or oceanographers by training. The human geographers are even more diverse as they include economists, philosophers, political scientists, social anthropologists, sociologists and so on. Even those trained as geographers may still have strong links with other disciplines, and indeed some stated that they could quite comfortably transfer to different departments, such as history, politics, sociology or social anthropology, although most (e.g. Jamuna, Rhine) did not want to because they felt that the geographical dimension was important, and not given enough weight in these other settings.

If, on the other hand, interdisciplinary research is regarded as the exploration of relations between the categories 'natural sciences', 'social sciences' and 'humanities', then the picture becomes a lot more complicated. The words of Rhine provide a good baseline for the discussion:

> People would normally think of interdisciplinary research as Geography and something else, but I consider research within Geography as interdisciplinary anyway, because the key boundary is between social and natural sciences. You can look at this as a boundary or you can look at it as a relationship, if you look at how a lot of social thought has changed over the last 15–20 years, you can see it has developed relationally, so you can see that the social and the natural are mutually constituted, rather than two separate realms between which there is a boundary which you occasionally cross ... the more we're pushed into the social construction of the biological, or the recognition of the biological basis of the social, then the more it is difficult to maintain this boundary ... In the last few years there has been a lot of emphasis in social science on notions of performance and practice and what people actually do, the embodiment, the embodied character of social life, and the embodied character of natural things, looking at work in the performing arts, dance and dramatology, looking at how you use those in social sciences and increasingly across the social and natural links ...

Opinions and inclinations

Although Rhine's opinion was that there are conceptual links between the natural and social sciences and the humanities, it was clear, nonetheless, that few had yet thought about this in any theoretically explicit or thorough manner. The few that had did not all necessarily agree with Rhine's opinion. For example, Mississippi opined that 'No vast intellectual project holds the discipline together', but followed this up with the point that in terms of research funding and resource flows, it is important for human and physical geographers to ally themselves together. The next paragraphs attempt to summarise the attitudes and the opinions of the respondents towards interdisciplinary research, and possible reasons for these. However, it must be borne in mind that the sample size was small, and our conclusions must therefore not be assumed to be representative.

The physical geographers were quite receptive to the idea of combining the approaches of the social and natural sciences, but they were generally thinking in terms of research with a practical implication, carried out to benefit society (e.g. Danube). Areas within physical geography where there seemed to be agreement that interdisciplinary approaches were most appropriate were: environmental management of pollution; the use of remote sensing or Geographical Information Systems (GIS) in resource mapping or in understanding the nature of a resource and its context in the development of the management of this resource; the application of earth sciences to the understanding of landslides to improve the management of human responses or to determine how different societies should plan to respond to this risk; the multiple views of scientists and scientific policymakers in relation to a particular project; behavioural issues related to how the environment is used; and agriculture and the use of indigenous knowledge in development.

Social scientists with an interest in development also had a practical understanding of why it was useful to combine the natural and the social sciences. Euphrates stated that '…by [interdisciplinary research's] very nature you end up dealing with "more real" issues'. Ganges explained that when working in marginal environments you had to have an understanding of both natural and social causes of problems. For example, even if you were doing plain social science research, you might still need to understand the nature of soils in a particular environment (anthropologists share the same view, as they realise that development must have a holistic approach).

Some of the human geographers appeared sceptical about the real possibilities of linking physical and human geography. Jamuna explained that there are two different models of geography:

> One [model] says that geography historically has attempted to combine two fundamentally very different approaches to research, natural and social sciences, and that in reality geography would be much more comfortable if the two sides went their separate ways … Another model is a unified vision of geography, with geography as an integrated subject with a historical tradition that has something to offer that the two sides on their own wouldn't. I am somewhat sceptical about this, because research has specialised so much, with the majority of research projects there isn't much benefit of being part of an integrated discipline called geography, they would have been fine on their own … The area that is usually mentioned as interdisciplinary is the environmental area, but I have always taken the view that this is a bit of a myth, that the environment is holding geography together, because people who study it, either study it as environmental scientists (so as physical geographers) or as social scientists … Only a small minority of people are genuinely doing both, doing both the environmental science stuff and the social science perspective.

These ideas find echo elsewhere: in discussing the move away from having a single human/physical tutor for the undergraduates as a result of the increase in specialisation, Amu Darya commented: 'I am surprised the subject is still holding together.'

There are two difficulties in making a link between the natural and the social. One lies in the conceptual and theoretical clash. According to Po, natural scientists are positivists, 'they all speak the same atomic language', thus they have a common currency. By contrast, social scientists privilege theory (Yangtze). Yangtze observed that 'post-structuralists have difficulties with something like soil science – it is a theoretical clash', but went on to express the opinion that 'in reality, the positivist/post-structuralist area is very fruitful to work in'.

The second difficulty is that over time, training in geography has changed. So, the difficulty experienced by some in linking the natural and the social could be generational.

Missouri said: 'The old school geographer had a very broad background, but they are disappearing as they get older. The younger colleagues tend to be a lot more specialised, and so in a sense, narrower.' Amazon, a younger colleague, backed this up by giving the example of a senior colleague who had made 'generalism his specialism' and had been able to occupy the common ground shared by physical and human geography, but described him as 'an increasingly rare beast'. Ganges explained that in a 'traditional' geography degree you would have studied both human and physical geography, which made you a 'real' geographer. Those who had only been exposed to human or physical geography did not qualify as 'real' geographers.

So, a number of respondents did not see interdisciplinary research as anything new, at least in the geographical context (Rhine, Paraná, Missouri, Yangtze). Geographers, even though they tend to specialise in either physical or human geography, have been exposed to both sides, so they have 'split personalities' (Paraná). They may be 'schizophrenic, but this is an asset because they can see what is on the other side of the boundary and they do not perceive it as threatening' (Paraná). Rhine stated that geography had always done interdisciplinary research; Rhine had always done interdisciplinary research. This sentiment was echoed by quite a few others: they had been trained in interdisciplinary approaches, so they saw no difficulties in combining different viewpoints or methodologies. It is curious that these contradicting viewpoints go largely unacknowledged by members of staff. It is almost as if staff do not wish to admit the existence of a marked divide between human and physical geography, a divide that was nevertheless noticed by undergraduate students as it manifested itself even during lectures, in sardonic allusions by both physical and human geographers to the comparative dullness of the 'other' side of geography. This divide could at least in part stem from the heated debate in the literature, and within and between geography departments, as to what geography really is, what it has to offer and where it is going in the future.

There exist varying degrees of allegiance towards geography as a discipline at Durham. There are those who view geography in a rather negative light. Mississippi stated that 'Some people go moist eyed and weepy about geography as a discipline but for me they are all tweed and beard. It's rather sad', and concluded that there was no justification for geography on the intellectual level. There were two coherent definitions of geography, one to do with 'where things are and how they got there', which was 'dull beyond words', and the other to do with 'the orchestration of processes in space and time', which did not justify a distinct discipline. However, in practice the existence of geography as a discipline could be justified because other disciplines simply ignored the orchestration of processes in space and time, whereas geography allowed the importance of space and time to be reasserted. Others also felt the lack of a coherent justification holding geography together. Jamuna explained: 'I have affection for geography because I've studied it all my life, but I don't make a fetish out of geography for the sake of geography' (however, the statement was qualified by making it clear that the issue whether geography should be a unitary discipline depends on whether one is discussing this matter at undergraduate or research level).

Paraná's position was slightly different, coming from an advocate for physical and human geography remaining within one department: 'Staying together keeps departments going because they are constantly prodding each other, and it is a process of cross-fertilisation. Evolution shows that mixing genes makes the product stronger.' Rhine also felt confident about geography as a discipline, and felt along with other colleagues that a strong case could be made as to what was intellectually distinctive about it, as it examined relations between people, nature and space.

Summary

In conclusion, despite all the generalised positive responses to the idea of interdisciplinary work, this optimism should not be taken at face value. For one thing, respondents did not always share a common definition or understanding of interdisciplinary research, and for another, they had individual definitions or understandings of what geography encompasses. Thus, it can be seen that not only is the definition of interdisciplinarity dependent on positionality, but the definition of geography is equally dependent on the same.

Commentary

Isabel Fletcher

This piece is structured around my route to becoming an interdisciplinary and/or transdisciplinary researcher. Woven into this narrative are explanations of what I found useful or thought-provoking about each of the four sets of excerpts included in this chapter. Much of this reflection involves using these texts to look back at my career because, as I explain below, at the time I was not aware of most of the academic literature on interdisciplinary research (and had never heard the term 'transdisciplinary'), and nor did I think of myself as an interdisciplinary researcher. Instead, I thought of myself as someone who was interested in interdisciplinary topics, such as how to feed the planet and how we understand the relationship between diet and health.

I have an undergraduate degree from the Open University (OU) that I did part-time while working in a range of jobs in the catering industry. The OU course requirements meant that I could study what interested me, and so I took courses across a range of disciplines, including cultural studies, development studies, English literature, gender studies, history of medicine and sociology. I graduated with what I now realise is, in a UK context, an unusually interdisciplinary undergraduate degree – in terms of topics of study and perhaps conceptual frameworks, if not research methods.

I returned to full-time study to undertake a Master's and then a PhD in science and technology studies (STS) at the University of Edinburgh. STS is generally accepted to have developed from the 1970s as an explicitly interdisciplinary field of research analysing science and technology using approaches derived from older disciplines such as anthropology, history, philosophy, political science and sociology. In my doctoral research I combined approaches from STS with those from history of medicine and policy studies. This might be considered a form of interdisciplinarity, but felt very much part of an existing STS approach that encouraged theoretical and methodological pluralism – a form of 'bricolage' where the researcher assembles the appropriate resources to understand a particular research topic.

At the time I was a PhD student, both European and national funders (such as the Economic and Social Research Council [ESRC] that funded my PhD) were encouraging researchers to undertake interdisciplinary research, especially on complex and policy-relevant social issues – what we would now call challenge- or mission-oriented research – and to engage with those in other sectors – usually described as 'stakeholders' – as part of transdisciplinary research processes. However, despite this encouragement, it was not made clear to us PhD students what interdisciplinary research involved and, in particular, how it was different from the mono-disciplinary model of research that structures universities and other academic institutions. This is one reason why I find Julie Thompson Klein's pioneering work on models of inter- and transdisciplinarity so valuable.[2]

Klein's typology of changing definitions of terms such as 'interdisciplinary' and 'transdisciplinary' provides an overview of the differing ways in which collaborative research has been conceptualised, highlighting the diverse communities of practice that have developed these definitions. This clarity is important because – as we found in the SHAPE-ID research (Vienni-Baptista et al, 2020) – shared definitions of these terms are often taken for granted. This is particularly true of the research policy literature where influential arguments about why and how to conduct such research circulate. This lack of clarity is compounded by the disparate and disjointed nature of the academic literatures on inter- and transdisciplinary and other forms of collaborative research: several different communities of practice have developed their own distinct bodies of literature analysing how to best conduct different forms of collaborative research. These literatures are not well connected to each other or the research policy literature. This causes problems – particularly for newcomers, such as PhD students and early career researchers – because it leads to confusion about what different terms mean in practice: what do interdisciplinary research, transdisciplinary research or team science involve, and how are they different from each other? It also leads to a cycle of 'reinventing the wheel' where the same research problems are rehearsed in the literature without acknowledging that solutions exist in other literatures (Vienni-Baptista et al, 2022).

This piece of Klein's writing (Extract 1.1) is a high-level overview of the topic. In contrast, the final excerpt by Lisa Lau and Margaret Pasquini (Extract 1.4) explores the attitudes of a group of geography researchers to interdisciplinarity, and here I see parallels between their accounts and my experience of being trained in STS (I am still uncertain whether it can or should be called a discipline). Recently I discovered the concept of a 'portmanteau' discipline (Lyall, 2019: 43), and I think this term captures what is important about fields of research like geography and STS, which have relatively porous boundaries and within which methodological pluralism is accepted or even encouraged. Extract 1.4 also highlights how the internal structure of disciplines influences the ways in which researchers working within it practise interdisciplinarity – for example, do you prioritise partnerships with other geographers, or look outside the discipline for potential collaborators? It shows how analysing interdisciplinarity and the ways in which we practise it entails thinking more clearly about what constitutes a discipline and how individual disciplines differ. Lau and Pasquini's material also demonstrates the different ways in which individual researchers identify (or not) with a discipline, something that these results show is as variable for these individuals as their understandings of interdisciplinarity.

The article by Philip Lowe and his co-authors (Extract 1.3) deals with the empirical, but in a different manner as it uses the Rural Economy and Land Use (Relu) project as a case study to explore why and how to undertake such research. Although the authors label it as interdisciplinary, Relu involved extensive interactions with a range of stakeholders, and now we might describe it as transdisciplinary research. In the excerpt we have selected, Lowe and his colleagues outline some of the reasons why social scientists might want to collaborate with natural scientists. They argue that joint problem framing between these groups was 'critical to developing projects that would approach a key question or set of issues from multiple angles, ensuring a more coherent set of solutions could be delivered' (Extract 1.3). This is very much how I was trained to undertake research, and has become an underlying principle in much policy-oriented research conducted in the UK and elsewhere.

At the same time, Lowe et al also describe some of the processes of collaborative research developed by project members, making the research seem more concrete and therefore achievable for novices. These processes were built into the structure of the project and, the authors argue, are necessary to achieve impactful social science. Refreshingly, despite the success of the Relu project, the authors carefully acknowledge the extra work that collaborative research involved, highlighting some of the main challenges that they encountered. These included differences between quantitative and qualitative research methods, competition between closely related disciplines and contrasting approaches to reflexivity and social critique, all of which I have experienced at some point in my postdoctoral career. When I finally read this piece, I found it reassuring to know that others, including established researchers, also experienced these issues and found them challenging.

In my postdoctoral career I have worked on a range of topics – always on fixed-term contracts and often for interdisciplinary centres and projects – as well as co-convening a cross-disciplinary network for researchers working on food-related topics (one of my central research interests). This has involved working with other social scientists as well as researchers from the humanities and biomedicine. Some of these projects have involved engagement across disciplines where we learned from each other's perspectives, but, in my experience, this has been rare. More common is an instrumental approach to collaboration where one discipline frames the research project and other disciplines are brought in to address particular research questions and often to undertake specific pre-assigned tasks. The excerpt by Andrew Barry and his co-authors (Extract 1.2) resonates strongly with this postdoctoral experience, and gave me a language to describe my interactions with other disciplines as well as those of other academics in my department who were conducting very similar studies to those described in the case studies. Barry and his colleagues describe three modes (or ideal-types) of collaboration: integrative-synthesis, which was seen as what we should aspire to; subordination-service was something as social scientists aiming to study science and scientists we were warned about (but did not necessarily have the capacity to change on our own); and finally the agonist-antagonist mode was not usually encouraged in the parts of STS that emphasised pragmatic engagement with science and science policy.

I have most often experienced subordination-service modes of interdisciplinarity – not just with colleagues from the natural or biomedical sciences, but also those from other more quantitative social sciences such as economics. This mode of engagement seems particularly common among quantitative researchers working in areas such as sustainability, especially if their work is closely aligned to the natural sciences. Funding agencies encourage large interdisciplinary proposals on high-profile topics such as improving food systems or combating anti-microbial resistance, with the stipulation that project teams include a range of disciplines. However, such encouragement does not always involve providing incentives for all the chosen disciplines to be involved in the important early stages of developing a proposal where the approach to the research is mapped out and the specific research questions are framed. In my experience, this leads to the predominance of subordination-service models of research, where qualitative researchers are still restricted to working on pre-defined topics such as public acceptability, attitudes or ethical implications of new technologies.

Elsewhere (Fletcher and Lyall, 2021), we have argued that it is possible for social scientists to undertake good quality research in such situations. Felicity Callard and Des Fitzgerald (see Extract 3.1) also reflect on this issue and come to a similar conclusion, arguing that integration is an idealised pipe dream and that we need to take the opportunities on offer from more limited roles or engagement. However, I want to retain the option of greater integration, particularly in the case of challenge-based research. Projects led by natural scientists often frame problems in restricted and often quite technical ways – How do we produce more food? How do we get people to eat better? How do we prevent the 'misuse' of antibiotics? – that unhelpfully restricts the kinds of research questions that can be asked, and therefore undermines the potential of inter- and transdisciplinary research to provide useful knowledge.

Despite a career that has involved working in various interdisciplinary contexts and a long-standing interest in topics that cross discipline boundaries, I do not consider myself an expert in inter- or transdisciplinary research. Only recently, and partly as a result of my participation in the SHAPE-ID project, have I begun to label myself as an interdisciplinary or even transdisciplinary researcher. Reflecting on this reluctance as I write this commentary, I conclude that a large part of it comes from a lack of knowledge – important-sounding pronouncements about the need for collaborative research to solve pressing social problems combined with a lack of information about how to go about these kinds of research made it seem unachievable. The key thing I learned from my participation in the SHAPE-ID project is that there are many different ways of undertaking collaborative research, and that there are different forms of knowledge involved – from the academic specialism of 'research on research' to the mundane knowledge of inclusive daily work practices and the importance of record keeping.

Notes

[1] Sadly, Julie Thompson Klein passed away in January 2023 during the production of this book. She is a greatly missed friend and mentor to many within the interdisciplinary and transdisciplinary research communities.

[2] For an STS researcher whose work focuses on interactions between science and policy, Klein's work also provides a valuable historical perspective on how models of science policy developed in the second half of the 20th century.

References and further reading

Fletcher, I. and Lyall, C. (2021) 'Stem Cells and Serendipity: Unburdening Social Scientists' Feelings of Failure', in D. Fam and M. O'Rourke (eds) *Interdisciplinary and Transdisciplinary Failures: Lessons Learned from Cautionary Tales*, Abingdon: Routledge, Chapter 3.

Lyall, C. (2019) *Being An Interdisciplinary Academic: How Institutions Shape University Careers*, London: Palgrave Pivot.

National Academy of Sciences, National Academy of Engineering and Institute of Medicine (2005) *Facilitating Interdisciplinary Research*, Washington, DC: The National Academies Press.

Pohl, C. and Hirsch Hadorn, G. (2008) 'Core Terms in Transdisciplinary Research', in G. Hirsch Hadorn, H. Hoffmann-Riem, S. Biber-Klemm, W. Grossenbacher-Mansuy, D. Joye, C. Pohl, U. Wiesmann and E. Zemp (eds) *Handbook of Transdisciplinary Research*, Berlin: Springer Dordrecht, pp 427–32.

Vienni-Baptista, B., Fletcher, I., Lyall, C. and Pohl, C. (2022) 'Embracing heterogeneity: Why plural understandings strengthen interdisciplinarity and transdisciplinarity', *Science and Public Policy*, 49(6): 865–77, https://doi.org/10.1093/scipol/scac034

Vienni-Baptista, B., Fletcher, I., Maryl, M., Wciślik, P., Buchner, A., Lyall, C., Spaapen, J. and Pohl, C. (2020) *Final Report on Understandings of Interdisciplinary and Transdisciplinary Research and Factors of Success and Failure*, Zenodo, https://doi.org/10.5281/zenodo.3824839

2

Developing Collaborative Conditions

Chapter overview

Collaboration is at the centre of interdisciplinarity and transdisciplinarity. As a key process in research, it entails making decisions on the reasons why the collaboration is needed, with whom to work, and what we want to achieve with the collaboration (see, for instance, the SHAPE-ID Toolkit 'Top Ten Tips' for a complete list of aspects to take into account before collaborating[1]). Collaboration needs dedication and commitment. It takes effort to build a successful team and to establish effective working relationships when the group consists of individuals coming from different backgrounds, working cultures and knowledge traditions. Researchers also face the challenge of integration (see Chapter 3 for a selection of readings on this topic).

The first reading from Catherine Lyall et al (Extract 2.1) offers practical guidelines for researchers and research managers who are seeking to develop interdisciplinary research strategies at a personal, institutional and multi-institutional level. The extract draws on examples from across the social and natural sciences, offering lessons, as well as tips for designing projects and managing teams. We have chosen material from the boxes and end of chapter questionnaires, as these are valuable resources for newcomers.

Taking the time to develop shared vocabularies and understandings in order to produce effective settings for inter- or transdisciplinary collaboration is a crucial task. In Extract 2.2, Louise Bracken and Elizabeth Oughton argue that constructing a common language can help in developing relationships of trust that will facilitate research between different disciplines. In the selected sections, they elaborate on how to bridge disciplinary differences using linguistic resources such as dialects or metaphors.

Researchers may also face the challenge of co-producing knowledge with different societal actors with divergent thought collectives: 'Knowledge coproduction is a particular description of a collaborative research', Christian Pohl and Gabriela Wülser argue in Extract 2.3. They introduce the main concepts in knowledge in co-production while detailing the vision behind the td-net toolbox developed by the Network of Transdisciplinary Research (td-net, Swiss Academies of Arts and Sciences, Switzerland). This provides hints on how to initiate, improve and profit from transdisciplinary research projects.

To deep-dive into a specific type of interdisciplinary collaboration between art and science, in Extract 2.4 James Leach presents some insights on an empirical study. He draws on the different worldviews that artists and scientists may bring to an interdisciplinary collaboration, and how these influence the potential results. These differences, and even contradictory perspectives, may enrich the collaboration and lead to better integration of the arts in interdisciplinary research.

EXTRACT 2.1

Lyall, C., Bruce, A., Tait, J. and Meagher, L. (2011) 'Making the Expedition a Success: Managing Interdisciplinary Projects and Teams', in *Interdisciplinary Research Journeys: Practical Strategies for Capturing Creativity*, London: Bloomsbury Academic, Chapter 4.

Key Advice 4.4 Tips for interdisciplinary team managers

Conceptualizing the research problem
- ensure that all participants contribute, and contribute to the same standard, even if their methodologies and data differ;
- negotiate roles as necessary; take the time to find a common framework for the research in order to get the right balance of contributions from the component disciplines (so that you achieve a truly interdisciplinary product, not simply a multidisciplinary project);
- plan to take extra time for group working in the early framing stages; facilitate lively interactions that help disciplinary partners explore commonalities and differences, and establish relationships and trust.

Distributing team responsibilities
- develop a systemic framework and agree the common problems and goals from the outset;
- build bridges between the different disciplinary contributions to achieve synergies across disciplines and methods;
- recognize that, despite early planning, interdisciplinary projects may need to develop and change as they proceed – be flexible;

- encourage the research team to be more reflective than they would be for a monodisciplinary project; facilitate frequent, open and positive discussion of the project's evolution;
- designate a dedicated member of staff to carry out coordination, dissemination and knowledge exchange responsibilities, as when such individuals are valued appropriately they can have a significant impact upon success of large-scale projects, programmes or centres;
- when distributing team responsibilities, be transparent so that every member of the team knows who will be doing what, and when;
- identify expertise and assign it appropriately without necessarily expecting everyone to participate fully in all tasks;
- be open to new methods;
- consider how analyses may be structured to integrate different sorts of findings, from different disciplines' methods and data;
- recognize that team responsibilities may go beyond standard/traditional areas of expertise;
- consider the role and contribution of 'users' or other stakeholders in the team.

Overcoming communication barriers
- expect to expend time and effort in developing a common language within the team;
- be aware that different disciplines have different traditions and styles of working; air preconceptions among partners about different disciplinary paradigms;
- include multiple face-to-face team-working meetings and networking events, especially early in the project and then at project milestones/decision-making points. Augment (but do not replace) as necessary with regular video-conferencing to tackle geographical separation;
- use social events to help the team coalesce. Joint fieldwork may also be helpful;
- find a way of applying rewards and incentives to teams rather than individuals;
- consider using existing techniques and computational tools for integrating data;
- provide opportunities for team members to write together to encourage integration across disciplines;
- expect some clashes within the team; when possible turn these into new ways of thinking about a research problem, or even new avenues for future research;
- steadfastly and diplomatically, throughout the project, refresh team members' commitment to their shared goals.

Bringing it all together
- as early as the research planning stage, consider how work – and credit – can be apportioned fairly when it comes time to publish results;
- be aware that different disciplines have different traditions in, for example, the sensitive issue of authorship which can sometimes be particularly disadvantageous to junior researchers within an interdisciplinary team;

- discuss a deliberate publications strategy with team members early on, toward the development of a portfolio of publications with different outputs targeted at different types of journals (e.g. various monodisciplinary journals as well as one or more interdisciplinary journals) or other media;
- designate lead responsibility for different publications to different team members depending on their disciplinary standing and their role within the team;
- before you reach the end of the project, you may also want to consider what factors will influence the likelihood of the team staying together and perhaps evolving (by adding or subtracting members to tackle what may be new niche opportunities building on the first round of research) or whether the team should be allowed to disperse and the project die a graceful death.

Continuing the journey

Shaping the nature of the endpoint is another challenge for managers of interdisciplinary projects. Usually, not all the subtleties and complexities of effective interdisciplinary projects will be completely resolved. However, funding stops at a certain point and individual researchers have to get on with their professional lives, securing other funding and/or pursuing somewhat different research problems. The end of an interdisciplinary project might consist of:

1   the interdisciplinary team dispersing completely (whether or not members have successfully generated integrated outcomes and publications from their work) with only a few individuals – or no one – pursuing the general problem area;
2   subsets of members of the team, perhaps augmented by new colleagues, pursuing subsets of the original problem (these new more focused teams might be monodisciplinary or interdisciplinary);
3   the original team, recognizable but with natural turnover, pursuing either the original problem or a next evolutionary stage/offshoot of that problem.

If sustainability of the investigation is desired – if some or all of the members of the expedition want the journey to continue – the team needs to be managed accordingly, long before the funding is over. For instance, team members will need to feel that their problem area continues to be vital and intellectually stimulating, offering new opportunities even beyond the end of a particular grant. The sorts of brainstorming activities described earlier (Key Advice 4.2), as well as continued networking and team-building, thus need to contribute continuously to the identification of compelling new questions derived in full or in part from work being done. Reflection and (self)-evaluation can help the team both to learn continuously about how to behave more effectively in an interdisciplinary context and also to monitor how their problem area is evolving generally, and where corresponding niches of opportunity may arise (Meagher and Lyall 2005b). Chapter 7 discusses various sorts of evaluation, including self-evaluation as a useful tool.

Leaders and coordinators may have special roles in 'succession planning', bringing along others to lead all or some components of the next phase of the work. Capacity-building in junior researchers who can grow to take on more responsible roles, and networking which can bring in new members with different perspectives, also contribute to a healthy evolution. Stokols and colleagues indicate positive influences that successful outcomes of one interdisciplinary study can have on future collaborative processes, through, for example, team members feeling intellectual satisfaction or an institution supporting future initiatives more effectively (by, for example, providing appropriate shared space) (Stokols et al. 2005). It may well be that the selection of leaders of new initiatives could favour those who have demonstrated requisite personal traits and prior experience of responsible roles in complex initiatives; innovative training programmes or indeed novel targeted training toolkits, can enhance the 'collaboration readiness' of future participants, helping to prepare a new cadre of individuals for roles in such initiatives (e.g. Stokols et al. 2008b; Mitrany and Stokols 2005; Nash 2008). Rhoten offers a compelling argument that mobility may not always be detrimental to the livelihood of interdisciplinary centres:

> while longer organizational life cycles give centers time to improve their research practices and processes, long-term and full-time affiliations can actually limit and not accentuate researcher creativity and productivity ... researchers who felt free to enter and exit collaborative relationships reported more progress with their interdisciplinary projects and greater satisfaction in their professional lives overall. (Rhoten 2004)

In going forward with follow-on proposals or initiatives, it makes sense to balance old and new, both in terms of drawing on findings from the earlier work to propose innovative research and in terms of continuing to involve some familiar faces, with trust already built, along with new ones.

[...]

Questions

For researchers
1   Interdisciplinary collaborations are often put together under a great deal of time pressure and may be conducted by people who do not know each other (or each others' disciplines) well. What are the key management issues that you need to address when working in these less than ideal circumstances?

For research managers
1   What steps (and in what order) need to be taken to proactively build and manage a team so that it makes the most of the potential value of interdisciplinarity?
2   What roles do you see yourself playing? How will you manage wearing multiple hats? Can you get support to help with any of your roles?

3   What role models do you have for successful interdisciplinary research? Can you find other managers of interdisciplinary initiatives with whom you can discuss issues?

4   In what way does the labelling of research team members in a collaboration (e.g. Principal Investigator, Co-investigator, etc.) influence their role in the team and their responsibilities and benefits institutionally and how might that affect specifically interdisciplinary projects?

For institutional leaders

1   What kind of job security do interdisciplinary researchers have in your organization? How does that compare with disciplinary experts?

2   How are 'teams' rewarded for research – or is all the reward on the basis of individuals?

3   How might this policy impact on interdisciplinary research?

4   Has your organization established any 'environments' conducive to interdisciplinarity?

For research funders

1   How would you evaluate the quality of a research team and the Principal Investigator's proposed management approach for an interdisciplinary research project?

2   Would you consider offering seedcorn funding to launch projects in new interdisciplinary directions?

3   Have you considered how self-evaluation or critical friend formative evaluations could help complex interdisciplinary projects evolve?

4   Would you consider bringing together interdisciplinary researchers to share experiences, approaches, issues and good practice regarding the management of interdisciplinary projects?

EXTRACT 2.2

Bracken, L.J. and Oughton, E.A. (2006) '"What do you mean?" The importance of language in developing interdisciplinary research', *Transactions of the Institute of British Geographers*, 31(3): 371–82.

Bridging disciplinary divides with language

Before explicitly examining the role of language, we explore some of the difficulties we noted in the Introduction on the role that language may play in contributing to the division. Distinctions between specialisms are striking in terms of their epistemologies: the ways in which they develop research questions and the methodologies chosen to explore those questions. Interdisciplinary research requires an understanding of the disciplines themselves as well as an understanding of how to connect disciplinary knowledge (Karlqvist 1999); methods and practice can therefore result in a barrier to this type of collaborative research. A simplistic view suggests that typically physical scientists treat the topic of study as an object, whereas to the social scientist the topic of study is the subject. As a consequence, physical scientists generally use methods to monitor and evaluate the object, whereas social scientists adopt methods that include reflection on their own role and effect on the research subject. This in turn leads to different writing styles, which can present difficulties for reporting discussion and results. For example, in physical sciences the use of the first person is rare, and writing distances the researcher from the object of research, whereas in social sciences the first person is used as a means of acknowledging the role and responsibility of the investigator.

The scale of research may vary between disciplines, which can make it difficult to relate knowledges. Stereotypically, Dalgaard et al. (2003) argue that social science disciplines are often labelled as 'soft' and tend to work at more regional levels, whereas physical sciences and the 'harder' disciplines tend to work at smaller scales. However, both physical and social sciences work on a range of spatial scales from the plot/individual to the global. Similarly, they operate on varying temporal scales. In contrast, Massey (1999 2005) suggests that the concept of space-time provides opportunities for physical and human geographers to work together. However, spatial and temporal differences in scale may lead to different definitions, and result in gaps in information flows, and consequent misunderstandings. Scale variations produce very different starting points from which to view the environment and hence can lead to diverse research strategies. We have described Dalgaard as offering a stereotype of 'soft' and 'hard' to reflect differences between physical and social scientific approaches within geography. We disagree with the categorization of 'hard' and 'soft', but our own experience in presentations and conversation at a recent 'interdisciplinary' meeting suggests that these attitudes are still alive and kicking.

An important but less recognized aspect of interdisciplinary work stems from the attitudes and feelings of the co-researchers. Lack of respect between physical and social scientists is mentioned in published articles (e.g. Bruce et al. 2004; Evans and Marvin 2004). This in turn leads to interdisciplinary research being regarded as of lower status; and as a consequence professional status and promotion may be affected adversely (Brewer 1999). There are also problems about where to publish results of interdisciplinary research, since often the nature of written articles does not sit happily within discipline-specific journals. This is a problem perceived as being exacerbated by the Research Assessment Exercise in British universities. Complementarity and cooperation rather than competitiveness may help in overcoming the negative effects of interdisciplinarity, and shared language has an important role to play here. Once knowledges are seen as embedded in different cultural contexts and there is mutual respect between specialisms, important lessons can be learnt and a much more fruitful collaboration instigated. More than this, the success of interdisciplinary research depends on nourishment drawn from shared disciplinary competence (Hansson 1999), which should remove any hierarchical value between different subjects.

Having touched briefly on the areas in which language has a key role in developing good inter-disciplinary research, we now develop and explore this role in more detail. We are trying to improve our own practice in interdisciplinary work by understanding the ways in which we use language and how it may help or hinder our understanding of each other and our respective disciplines. Language is a living thing and evolves in everyday use; it also evolves in its use within disciplines. Referring back to dictionaries of human and physical geography shows how the emphasis in description of terms and the context for descriptions change, although noticeably less in the physical than the human references (Johnston et al. 1986 1994; Clark 1985). Subject dictionaries tend to be used to inform students and those outside the discipline of common understandings within. However, formal definitions fail to capture the breadth and dynamism of language in use. We are not attempting here a study of the changed meanings of commonly used terms within geography over the long term, but are concerned more with the contemporary, multiple meanings and uses of words in practice.

Language may determine the positionality of the researcher, the way in which the research question is framed, the translation of the 'field' to the academy and the development of the theoretical context (see, for example, Quinn and Holland 1987; Mirowski 1994; Pryke et al. 2003). Previously, Demeritt and Dyer (2002) have analysed the role of dialogue in geographical research, in particular the range of definitions of dialogue and how these map onto ongoing controversies within geography. However, within their paper there is little reference to everyday practice. In the course of our work there have been occasional but important instances where the meaning or intent of the speaker has not been understood by the listener. So far we have identified three distinct issues: *dialects, metaphor* and *articulation*. We will explore each of these within their ethnographic context as they appeared in the use of three words: *dynamic, mapping* and *catchment*.

Dialects

The first aspect of the use of language which we wish to highlight is what Wear (1999) refers to as *dialects*. *Dialects* represent the difference between everyday use of a word and expert use, and the ways in which different disciplines use the same word to mean different things. *Dialects* are also produced by the same word having slightly different meanings within different disciplines (Bruce et al. 2004), again different from the everyday meaning. Words which are in everyday use by non experts tend to be those that cause the most difficulty for the unwary practitioner. These misunderstandings may be exacerbated by the fact that academics, articulate by nature, are unlikely to question the meaning of a word with which they are already familiar. As Wear notes:

> Language is most important because scientists speak in dialects that are specialised to their disciplines. Unfortunately, these dialects can at times sound like common language, leading the uninitiated reader to the mistaken conclusion that she understands what is being said. (1999, 299)

The conversation may be well developed before it becomes apparent that a particular word has a specific disciplinary interpretation as well as its everyday use. This situation is bound to lead to frustration on both sides.

The word we wish to use as an example here is *dynamic*. The context is our visit to the River Esk. Our analysis is of a discussion about processes and methodology that took place in the field. The intention was for the social scientists to familiarize themselves with the location and for the physical scientist to explain her current research. This involved visiting a reach of the river where instrumentation was already in place and she described her methods and the ways in which her research related to the detailed processes of landscape evolution. The physical scientist described a process using simple scientific terms with the intention of giving a generic description of sediment transport. In a discussion of farmers' understanding of the landscape, the physical scientist described the catchment response to rainfall as being *dynamic*. She argued that farmers would understand the gross effects, but would not necessarily understand the detailed hydrological processes.

The social scientists as a group queried the method and context of the fluvial geomorphology research, for example, the choice of location of measuring equipment, changes in location, frequency of measurements, choice of scale and the boundaries of the research. They took for granted that farmers would understand the process through their experience of living on and working the land. This different understanding of the word *dynamic* led to a series of heated debates about different stakeholders' knowledge of the landscape. For a while we left the subject and moved on to other things. Once we had returned from the field, we (the authors) were reflecting on how difficult and interesting the day had been and started to work back from the disagreements over stakeholders to try and understand how our differences had arisen. We traced it back to the word *dynamic*.

Dynamic has both everyday meanings and discipline-specific meanings. As an adverb, the OED (1993) defines *dynamic* as 'of force in actual operation', and this was understood and implicitly used by both participants in the conversation. The problem lay in the differences in the perceived time and spatial scales to which *dynamic* referred between disciplinary and normal use. To the physical geographer, *dynamic* meant that stream discharge would be variable depending on the antecedent moisture conditions of the catchment over very short timescales of a few hours to a few days. The social scientists understood *dynamic* to mean relatively rapid changes over longer timescales, undefined. This confusion could easily have been clarified on the spot had we recognized this as a *dialect* word. The implications for planning the research in the field were huge, and snowballed from a very simple misunderstanding. This example shows how we got to very different endpoints from a poor matching of understanding of one word. In the company of experts of the same discipline, this misunderstanding would (probably) not have happened.

Metaphor

The second aspect of language is *metaphor*. At the simplest level, metaphors clarify and illuminate an argument and are commonly used to assist in teaching. At a more complex level, Klamor and Leonard (1994) discuss 'heuristic metaphor', meaning a metaphor that develops thinking in a new direction, and which is open to further development in a systematic manner through further analogy. This allows us to understand in ways that 'a literal rendering cannot'. Heuristic metaphor is commonly drawn on by specialists in talking to each other. The significant aspect of heuristic metaphor is that the metaphor itself does not say, it suggests. This implies that those who do not share the context do not necessarily interpret the metaphor.

> Experimentally, the problem of interpretation is illustrated by presenting even a common metaphor to young children or to anyone likewise removed from your 'speech community'. (Klamor and Leonard 1994, 29)

The significance of 'speech community' for communication between disciplines is well illustrated by the following reference to the economists' use of language:

> economists, as well as those in other fields, communicate mainly with powerful figures of speech – in particular metaphors and appeals to authority – that offer up a compact and rich way of communicating within a peer group (even when these figures are enacted without a full understanding of their content). They also have the effect of excluding others from the conversation. (McCloskey in Wear 1999, 299)

At a third level Klamor and Leonard refer to 'constitutive metaphor'; the metaphor interprets a world which is unknowable, 'or at least unknown'. The example they use is that of genes being represented by a code. In this case the metaphor may become so entrenched it is regarded as being true. All three types of metaphor are common in our everyday discourse. They are embedded in our language and we rarely think about them or are aware that we use them. For the most part we share the speech community. For good interdisciplinary practice we need to be aware of the times at which we move into separated speech communities and when the form of metaphor being used may be misinterpreted.

The metaphor that we draw on is *mapping*. The context of the discussion was the development of the analytical framework to underpin our study (Figure 1). Our understanding and use of metaphor did not bring about the type of discipline-related disagreement that *dialects* had generated. In this case, we were conscious of the role that *mapping* played as a metaphor. We did not intend to use the term in the sense of either 'relational hieroglyphics to represent the landscape' or as a diagrammatic systems framework as used by both physical and human geographers. We were seeking a framework that would allow us to relate differently conceived social relations and embed them within the physical landscape, to explore the complex interrelationships ongoing within rural areas between different groupings of humans, but also between humans and the landscape. We worked on this idea together and the metaphor *mapping* provided us with a name for this activity. In this way we could achieve agreement because we were using a relatively empty metaphor. This conjures up an image of a multilayered, complex web which is both affected by and affects everyday life. *Mapping* is the systematic description of the processes involved. This is also a very different interpretation of *mapping* from that used in everyday language, and is therefore a dialect word as well. Its use was clear to us because we developed it explicitly to meet our needs in this research project. However, were we to simply refer to our work in discussion with others as 'mapping catchment interactions', a completely different product could be imagined.

In this particular example, metaphor was a positive contribution to interdisciplinary development. However, further reflection revealed the very large number of specifically disciplinary metaphors that we draw upon and the potential pitfalls that they present. The process of developing our common understanding of metaphor draws us into our final area of language to be discussed and that is *articulation*.

Articulation

The term *articulation* is borrowed from Ramadier (2004). This aspect of language differs from the first two in that it is a process rather than a register of speech. *Articulation* as described by Ramadier involves deconstructing one's own disciplinary knowledge in conjunction with those of other disciplines in order to understand the building blocks and thereby reconstruct a common understanding. It should be made clear, however, that

> articulation is what enables us to seek coherence within paradoxes, and *not unity*. (Ramadier 2004, 432; our emphases)

We find this idea of *articulation* particularly stimulating, and an accurate description of the very active discussions that we had when we visited our field site.

The context for our discussion was the first meeting to discuss the aims of the land and water theme in the RELU programme. The physical scientist started to talk about her work in the Esk catchment. One of the social scientists asked what *catchment* meant. The physical scientist described a *catchment* as the area of land defined by the watershed (drainage boundaries) of a *particular* river. The social scientists were willing to accept this, but were concerned that it had little meaning as a boundary for social and economic processes related to the physical landscape. Furthermore, the economist and human geographer had different conceptions of what these might be. That is not to imply that the physical scientist didn't recognize that human activities affect the environment, but that she chose to discount these, since they are not seen as *immediately* relevant to the physical processes. This was an uncomfortable moment for the physical scientist, although we all appreciated that this type of challenge was an important part of articulation. It was in the process of building a common understanding of the word *catchment* that the physical scientist recognized the implications of limiting research to a narrow definition based solely on physical topography, i.e. looking at a landscape without including human beings.

These slightly different definitions of *catchment* also serve to highlight two alternative starting points in terms of thinking about the landscape, related to disciplinary backgrounds. The social scientists' premise for thinking about the landscape was based on the ways in which human beings interact with the environment and how this interaction and behaviour is then affected by environmental processes. On the other hand the physical scientists' premise was based on the river producing a signal which reflected the processes ongoing in the landscape. The river therefore responds to a series of inputs, primarily physical processes, although these can be subject to modification by human beings. Different disciplines therefore tend to have different starting points for thinking about the landscape, as well as being trained to think in different ways. This results in conceptual boundaries being drawn in very different ways and at different spatial and temporal scales. Using the process of articulation to deconstruct these together and in an interdisciplinary manner allowed us to produce a more complex definition of catchment and a much more powerful basis from which to formulate research questions.

This was an important moment. We felt that we had built up a much better understanding by clarifying, justifying and arguing. Through this process we moved much closer to the crux of the problem that we wished to explore and gained a deeper insight into what interdisciplinarity really meant. One aspect of this progress was to define a whole that was greater than the sum of its parts. A second aspect emerged through the process. Each member of the team was constantly tested in their assumptions and perceptions, and whilst the process was difficult and time consuming it was rewarding and resulted in a much stronger basis from which to develop research. It is here that the micro-politics of relationships become important. There is a potential pitfall here that we need to make explicit and that is one of trust. Intellectual egos may be fragile and within our group we recognized that we felt that we were taking risks. As we have pointed out, although we had not all worked together before, we did form a network of friends and did not feel vulnerable by exposing our lack of knowledge. It was important to be able to expose disciplinary ignorance, acknowledge weaknesses and build on strengths.

EXTRACT 2.3

Pohl, C. and Wülser, G. (2019) 'Methods for Coproduction of Knowledge Among Diverse Disciplines and Stakeholders', in K. Hall, A. Vogel and R. Croyle (eds) *Strategies for Team Science Success*, Cham, Switzerland: Springer, Chapter 8.

8.1. Introduction

In order to address complex issues—such as public health, migration, or sustainable development—in an encompassing way, knowledge of different fields is needed. This knowledge is provided, further developed, and transferred to the next generation by universities and their disciplines. Which disciplines should be involved in a project depends on the aims and scope of the project, for example, the scales that should be included in an investigation (e.g., molecules, cells, organs, individuals, specific societal groups, or groups of different nations and cultures). If the aim of research is to have an impact on how an issue is dealt with by society, concerned societal actors have to be involved in the research process, too. This is not only to learn about and take into account their stakes and the power relations between them; importantly societal actors might also have relevant expertise on the issue.

Researchers from different disciplines as well as societal actors enter such collaborations with diverse perspectives on the world. In the words of Fleck (1979), they are members of different thought collectives who share a specific thought style, that is, a specific way of looking at and making sense of the world. For instance, regarding the same piece of land, an oil company might see extractable and marketable oil, the

First Nation people a sacred place for their gods and ancestors, the engineer a chance to build a new extracting technology, an environmental NGO potential environmental damage, and a law researcher a case to study the rights of First Nation people. All these actors look at the same piece of land. However, as part of their professional or academic training, they have learned to focus on specific aspects of the land and to fade out other aspects. On the one hand, this specialization has provided us with an enormous amount of knowledge, technologies, and societal practices. On the other hand, it also bears the risk that complex issues are framed, analyzed, and solved using only one specific thought style, for instance, the most powerful one. Such a reduction of complexity does not only create tensions with the thought collectives that are less powerful but also runs the risk of providing partial solutions. Partial solutions solve some aspects of the issue (for instance, oil extraction) and, at the same time, create negative side effects (for instance, the destruction of sacred lands).

Scientific attempts to deal with societal issues in an encompassing way are labeled interdisciplinary or transdisciplinary. The call for such approaches has a long tradition (Winch 1947; Anonymous 1966; Jantsch 1970), often at a low level of intensity but with phases of higher interest (cf. Apostel 1972). For more than two decades, there has been continuous and steadily growing interest in interdisciplinary and transdisciplinary research (Klein 1996; Klein et al. 2001; NAS/NAE/IOM 2005; Hirsch Hadorn et al. 2008; Frodeman et al. 2010). In the field of health, the question of how to collaborate in teams has led to a completely new field of research: the Science of Team Science (Stokols et al. 2008; Falk-Krzesinski et al. 2011; National Research Council 2015).

A key question of interdisciplinarity and transdisciplinarity is how different thought collectives collaborate and integrate their thought styles. If members of different thought collectives collaborate on an issue, they can no longer perform their analysis in isolation. Integration means that the way in which they frame the issue, identify the main research questions, analyze the questions, and try to have an impact is discussed and deliberated among those who participate in the project team. This collaborative effort is called coproduction of knowledge (Lemos and Morehouse 2005; Robinson and Tansey 2006; Polk 2015). The prefix "co-" is used to denote that the production of knowledge is no longer done independently of other thought collectives and that it is influenced by them.

A number of scholars have recently started to develop and collect tools and methods that support team science or the coproduction of knowledge (McDonald et al. 2009; Bergmann et al. 2012; Vogel et al. 2013). They have also compiled online toolboxes, such as the Team Science Toolkit, Integration and Implementation Sciences tools, or td-net's toolbox for coproducing knowledge. In the following sections, we introduce some of the specific challenges of coproduction and give examples of how the tools can help overcome them. For that purpose, we elaborate on td-net's toolbox for coproducing knowledge, which we have been editing. First, we present a background on the main challenges that the toolbox addresses. We then give a short overview of the methods included in the toolbox and present three tools in more detail.

8.2. Brokering Images of Knowledge

Knowledge coproduction is a particular description of a collaborative research process. In the "continuing discussion about whether TD [transdisciplinarity] is descriptive of a research process or whether it best describes the research outcomes that eventually emerge from projects that may include some blend of MD [multidisciplinary], ID [interdisciplinary], and TD processes" (Stokols et al. 2013, p. 5), coproduction stands for first position. It describes the transdisciplinary process of framing and analyzing an issue and, when needed, to develop measures and an action plan.

In terms of knowledge, coproduction is an approach aimed at overcoming the "symmetry of ignorance" (Rittel 1984, p. 325). Each of the thought collectives of researchers (with their specialized disciplinary knowledge) and societal actors (with their expertise as members of civil society and the private and public sectors) provides only part of the knowledge that is needed to come to a comprehensive understanding and management of an issue. The knowledge generated by each thought collective is embedded in a particular thought style, a certain way of perceiving an issue. If the thought styles that coproduce knowledge differ in how they perceive an issue, a feeling of misunderstanding each other and the need to clarify terms might evolve. This is because members of a thought collective do not only share a specific way of perceiving an issue; they also share basic assumptions about, for instance, what the right concepts for describing an issue are or what makes arguments about the issue trustworthy. In the words of Elkana (1979), this is because thought styles consist of both a body of knowledge (the knowledge itself) and an image of knowledge (ideas about what constitutes "good" knowledge and how it should be produced). Therefore, what in coproduction might first seem like a question of language and terminology might turn out be a question of the underlying images of knowledge. Moreover, what seems to require honest knowledge brokering (Pielke 2007) between thought collectives in fact requires the honest brokering of images of knowledge (Pohl 2011).

EXTRACT 2.4

Leach, J. (2011) 'The self of the scientist, material for the artist: Emergent distinctions in an interdisciplinary collaboration', *Social Analysis*, 55(3): 143–63.

The Subject Matter of Art and of Science

Art and science have a different focus in the understanding of my informants. As a physicist participating in the Scheme put it: "We [scientists] are unveiling bits of nature." His collaborator (a novelist) told me: "[My collaborating physicist] holds the idea that there is an objective reality we come along and discover, [whereas] I think that truth depends on there being language in order to know it." This points to the role that perceptions of subjectivity and objectivity had in the making of these different kinds of knowledge. Another enthusiastic scientist described the difficulty he had faced with the artist he had accepted into his laboratory over accuracy in representing the scientific facts in the artwork. For him, the issue was particularly pertinent because his enthusiasm over involvement in the Scheme was based on his desire to communicate the wonder of science. As he put it: "The artist may be seeking to allow an audience to interpret and question. Science as communication requires a more direct pedagogical approach." In another partnership, the artist confessed: "I don't understand everything by any means. This made me worried at first—very worried— but then I realized that I am an artist, and I don't have to understand everything. A different perspective is allowed. Misunderstanding allows me [a] different perspective." These contrasts point to the operational understanding that many of the scientists brought to the Scheme, that is, that they are forced to deal with what they find (as opposed to what they would like to find) or what they construct.

There was no denial that artifice was seen as being central to the knowledge-making practice of science, most fundamentally, with regard to the construction of large laboratories that had made scientific investigations possible. In the scientists' self-representations, the fact that science had become a large-scale operation meant that a division of labor was now necessary and that no one person could encompass all the expertise necessary for scientific discovery. That the results of investigations do not depend upon the context or the person of the investigator was an explicit part of the participating scientists' understanding.

This was not the case for the artists in the collaborations, who made their claims to the status of artist and to the ownership of their creations by presenting the material that they had worked on as internal, as it were, to their person. Let me explain. If sense is achieved by scientists because of regularities in the reality upon which they work that are external to themselves or their ability to make connections, as it were, the meaning of artwork is in the connections that artists are able to make internally within themselves, as perceiving and thinking subjects, and then express in their works of art. For the artists in the Scheme, sense was an expression of internal creativity—not external reality—and thus was not of the same type as that which the scientists attained. As another informant put it: "I characterized art as being concerned with individualism and self-expression, while science is driven mostly by curiosity about the world."

One can see the consistent construction of a difference, one that is found in the combination of the subject matter and the approach that is deemed suitable for presenting it. I highlight the insistence on, and institutionalized constitution of, an external world existent as a reality beyond any particular perceiver as the subject matter for science. As Law (2007: 599–601) has argued more widely, scientists assume that a thing like a habitat or ecosystem is a real entity found in nature: it has a particular "out-there-ness" that lends it singularity. Conversely, the subject matter for art is a subjective, interpretive connection; thus, it features an internal, rather than external, focus (whatever the actual subject of the artwork might be). Both have value, yet notions about the constitution of this differing value and the way in which it ought to be connected to the person of its producer have their foundation in these distinctions.

The distinction between disciplines was clearly a given for those sponsoring the Scheme. However, the material introducing it, as well as statements made by participants, demonstrated the perception that the arts and sciences should be alternatives within the same set of practices. "Can questions posed by scientists be posed by artists to achieve results in another medium?" was how one participant put it. So we can observe that the interface between the practices then must rest, conceptually, on some commonality and on some inherent difference. I believe that the commonality was made possible for the participants by a contemporary notion that conceives of 'knowledge' as intangible objects that can be externalized from their producers and that appear to carry their value despite this abstraction (see Leach 2012). That is, by describing the outcomes of scientific research and artistic practice as 'knowledge production', some level of commensurability was implied. That in turn prompted questions about how to combine them and which kind is most useful, most valuable, and so forth, in a resource-constrained environment. This move shifted the focus of attention from the relations of creation—including those between the persons (disciplines) themselves—to the objects produced. These objects contain alternative versions of 'valuable knowledge' and thus invoke intellectual property law as their background. This was underlined by the emergent distinctions that referred back to the separate logics of copyright and patent claims. However, to make hybrid objects containing the value of both was an aim of the Scheme. This goal was made possible by an understanding of the arts and the sciences as 'knowledge producing'.

In making claims to knowledge through intellectual property, aspects of the production process are explicitly or implicitly referred to in the evidence for the claim. The different kinds of connection that a maker has to the object produced appear extrinsic to its existence once it is separate and appears to carry its value as a 'knowledge object' without reference to its producer. Whoever originally owns the object is a retrospective reconstruction, as it were, of the processes of production. Differences in the processes of making are put down to the requirements of different kinds of subject matter (as will be discussed below) and in the location of the potential effects of the created object. For example, scientific knowledge of medicinal plants refers to existent entities given in nature, and having that knowledge has its primary effect in the bodies of patients, not in the world of culture and art. But of course, plant knowledge is already a cultural artifact, and cultural factors are also evident in how health and well-being are judged. The requirements for understanding and utilizing the kinds of knowledge made in art or science tend to overlap, and thus purification of both object and producer are demanded at different moments in each process in order for claims over the status of such productions to stand.

Commentary

Bianca Vienni-Baptista

Collaboration in inter- and transdisciplinary settings is a journey, as Catherine Lyall and co-authors explain in Extract 2.1. In such an adventure, conditions influencing the process of collaborating may change, even during the trip, and teams as individuals need to make sure that common aims are encountered. Readings collected in this chapter offer practical tips to cope with challenges when collaborating with different disciplines (Extracts 2.1 and 2.2), insights on how to co-create knowledge with the arts (Extract 2.4), and toolkits and methods to be applied in the process of integration (Extract 2.3). These extracts pitch strategies for researchers and managers alike.

Practising inter- and transdisciplinarity require acknowledging differences and working productively with and across them. In Extract 2.2, Louise Bracken and Elizabeth Oughton analyse the implications of such differences between the natural and social sciences. In my own experience, this constant negotiation that collaboration implies demands a trust-building process, in which commitment and caring are indispensable factors.

In the EU project SHAPE-ID (see the 'Introductory Essay' for further details), we sought to provide new insights and examples of good practice and tips on how to overcome difficulties in inter- and transdisciplinary research endeavours (Spaapen et al, 2020). In this commentary, I build on the findings of our survey to elaborate on connections with the readings in this chapter. In the survey, we identified a set of factors that influence collaboration, and compared them with the experiences of a smaller number of researchers working on large-scale European inter- and transdisciplinary projects in which arts, humanities and social sciences (AHSS) disciplines play a leading role (see the 'Introductory Essay' for more details) (Spaapen et al, 2020).

Together with my colleagues, we found factors related to academic cultures and epistemologies to be the most common obstacles in large inter- and transdisciplinary teams. In multinational consortia, such as the ones built in Horizon 2020 and Horizon Europe programmes, collaborations need to be explicitly planned and practised, as Catherine Lyall et al argue in Extract 2.1. This process that may begin with informal collaborations or elaborated ways of integration differ in purposes and scope, but depends on the disciplines, perspectives and personal skills of the scientific and societal actors that are involved in the research process (td-net, 2020).

Most projects participating in the survey were the result of long-standing – formal or informal – collaborations between partners that had exchanged knowledge and built trust among participants. When asked about the relationship with science, technology, engineering, mathematics and medicine (STEMM) partners, most respondents were positive about that kind of cooperation, although some researchers mentioned that differences were sometimes hard to bridge (Spaapen et al, 2020).

Constraints are everywhere, as these extracts show. Similarly to what Louise Bracken and Elizabeth Oughton argue in their article (Extract 2.2), in our survey we found out that researchers from AHSS are usually concerned with how STEMM researchers adopt and co-opt the vocabulary and methods developed by AHSS without integration taking place (Spaapen et al, 2020).

In other cases, AHSS disciplines, such as philosophy, position themselves as a provider of questions and new insights in inter- and transdisciplinary research. Problems associated with the assumptions that STEMM researchers have in relation to AHSS and their role in the research process were also seen as relevant according to EU researchers whom we surveyed. These assumptions can range from more romanticised visions to instrumental perspectives on AHSS roles in inter- and transdisciplinary research. Several extracts in this book confirm that a substantive obstacle in inter- and transdisciplinary research is to actually overcome the 'ignorance' between disciplines that try to jointly collaborate (see Chapters 1 and 4 in this book, for example).

Another example of constraints in collaboration follows observations made by James Leach (Extract 2.4) in relation to the role of the visual and performing arts in inter- and transdisciplinary research. These have not received much attention in the literature, but some references acknowledge that they are seen as using a different language to other fields in inter- or transdisciplinary settings. The question of how the problem is interpreted and framed is often posed to the audience in the arts, not given in advance, while most other researchers delineate a research question as the initial step in the research process (Rust, 2007). Leach rightly observes: 'the artists in the collaborations … made their claims to the status of artist and to the ownership of their creations by presenting the material that they had worked on as internal, as it were, to their person.'

Integration is one of the main processes we foster when working collaboratively. We aim to bring together different knowledge(s) and combine them in new ways, *integrating* them to approach multiple dimensions in complex problems. The root cause of difficulties regarding AHSS integration in inter- and transdisciplinary research is to be found in a lack of understanding among researchers, evaluators, policymakers and funders of what 'AHSS' is, and what the individual disciplines that are aggregated under this acronym can contribute to solving problems in society. Attitudinal change is necessary in all these sectors of the research system, so that funders who champion AHSS research integration do not have to counter prejudices and preconceptions before becoming effective (Spaapen et al, 2020; Vienni-Baptista et al, 2020).

Skills needed to foster better and more caring collaborative processes span from individual organisational aspects to responsibilities that account for the group wellbeing to accomplish shared goals. This does not come without challenges, as Catherine Lyall and co-authors indicate in Extract 2.1, and need to be accompanied by a variety of procedures to achieve both collaboration and communication between disciplines. These depend on the research questions and the goals of each project. Strategies to promote better collaboration and integration range from informal meetings, regular seminars, creation of further working groups, joint publications and more structured methodologies that are tailored to each project (Extract 2.3).

In Extract 2.3 Christian Pohl and Gabriela Wülser point out that:

a number of scholars have recently started to develop and collect tools and methods that support team science or the coproduction of knowledge (McDonald et al. 2009; Bergmann et al. 2012; Vogel et al. 2013). They have also compiled online toolboxes, such as the Team Science Toolkit, Integration and Implementation Sciences tools, or td-net's toolbox for coproducing knowledge.

Another example is the SHAPE-ID Toolkit,[2] built to foster new pathways between AHSS and STEMM in inter- and transdisciplinary research (see the 'Introductory Essay').

Measures for overcoming barriers or difficulties include strategies to arrive at a consensus within the group, methods found in the literature or innovative ones developed and implemented within a research team. All these aim at building common ground within the research team and fostering integration. Our survey confirmed that consensus among partners seems the most frequent strategy, reached by discussion giving all participants, including stakeholders, an equal voice. Extensive consultation complements this strategy (Spaapen et al, 2020).

Regarding new methods and tools developed to overcome these constraints, EU researchers shared the following recommendations (Spaapen et al, 2020):

- To develop an 'output matrix' to consolidate expectations (where members of the team could detail what do they expect from others, what can they contribute to others) (Survey 8);
- To hire an expert on transdisciplinary evaluation who accompanied the team from day 1 (Survey 8);
- To hire a creative facilitator (Survey 12);

- 'I like, I wish' methodology for facilitated feedback (Survey 6);
- Weekly community breakfast (Survey 6).

To conclude, recognition of the diverse factors that influence inter- and transdisciplinary research collaboration would help to manage a team and accomplish better research results. With the SHAPE-ID team, we argued that better pathways to collaboration need time and money to develop mutual understanding between potential partners, including funders, evaluators and the different stakeholder communities.

Given the diverse factors that may hinder inter- and transdisciplinary collaborations, support (with specific funding) is needed for the creation of toolkits that relate factors to actors (for example, researchers, funders and policymakers) able to influence them and guide research and policy processes alike (Vienni-Baptista et al, 2020). Recently, the SHAPE-ID team and members from the Working Group on Toolkits (International Global Alliance for Research and Education in Interdisciplinarity and Transdisciplinarity, ITD Alliance) face this pervasive challenge. Toolkits for supporting and enhancing inter- and transdisciplinary research are already available as online resources, but their sustainability in the long term is uncertain. These efforts are weak as teams do not have specific funding to assure their service to the scientific and practitioners' communities. Usually, after the funding period is finished, toolkits survive for a few years thanks to the voluntary work of committed teams, but tend to disappear, leaving a great amount of time and money wasted.

Toolkits may contribute to documenting and systematising the variety of research processes and practices of AHSS integration in inter- and transdisciplinary research. This would lower entry barriers and avoid research teams 'reinventing the wheel' each time an inter- or transdisciplinary project is developed. Toolkits also allow the current state of the art of inter- and transdisciplinary research to be recognised (Vienni-Baptista et al, 2020). From collaborative processes, innovative approaches to complex problems can emerge, implying new interactions between disciplines and greater integration of AHSS disciplines.

Notes
[1] www.shapeidtoolkit.eu/wp-content/uploads/2021/05/Top-ten-tips-evaluation.pdf
[2] www.shapeidtoolkit.eu

References and further reading
Bozeman, B., Gaughan, M., Youtie, J., Slade, C.P. and Rimes, H. (2016) 'Research collaboration experiences, good and bad: Dispatches from the front lines', *Science and Public Policy*, 43(2): 226–44, https://doi.org/10.1093/scipol/scv035

Hall, K., Vogel, A. and Croyle, R. (eds) (2019) *Strategies for Team Science Success*, Cham, Switzerland: Springer.

Klein, J.T. (1996) *Crossing Boundaries: Knowledge, Disciplinarities, and Interdisciplinarities*, Charlottesville, VA: University Press of Virginia.

Klein, J.T. (2021) *Beyond Interdisciplinarity: Boundary Work, Communication, and Collaboration*, New York and Oxford: Oxford University Press.

Muhonen, R., Benneworth, P. and Olmos-Peñuela, J. (2019) 'From productive interactions to impact pathways: Understanding the key dimensions in developing SSH research societal impact', *Research Evaluation*, 29(1): 34–47, https://doi.org/10.1093/reseval/rvz003

Prainsack, B., Svendsen, M.N., Kock, L. and Ehrich, K. (2010) 'How do we collaborate? Social science researchers' experience of interdisciplinarity in biomedical settings', *BioSocieties*, 5(2): 278–86, https://doi.org/10.1057/biosoc.2010.7

Rust, C. (2007) 'Unstated contributions: How artistic inquiry can inform inter-disciplinary research', *International Journal of Design*, 1(3): 69–76, http://shura.shu.ac.uk/966

Spaapen, J., Vienni-Baptista, B., Buchner, A. and Pohl, C. (2020) *Report on Survey among Interdisciplinary and Transdisciplinary Researchers and Post-Survey Interviews with Policy Stakeholders*, H2020 Project 'Shaping interdisciplinary practices in Europe', Zenodo, https://doi.org/10.5281/zenodo.3824727

td-net (Network for Transdisciplinary Research) (2020) td-net news. www.transdisciplinarity.ch/en/td-net/Aktuell/td-net-News.html

Vienni-Baptista, B., Lyall, C., Ohlmeyer, J., Spaapen, J., Wallace, D. and Pohl, C. (2020) *Improving Pathways to Interdisciplinary and Transdisciplinary Research for the Arts, Humanities and Social Sciences: First Lessons from the SHAPE-ID project – Policy Brief*, Zenodo, https://doi.org/10.5281/zenodo.3824954

3

Co-Creating a Research Project

Chapter overview

Increasingly, academic research and the broader generation and sharing of knowledge are taking place, not just across different disciplines within research-performing organisations but also among a much wider range of stakeholders who may have different forms of knowledge to offer. This requires both an acknowledgement and the active integration of different cultural and disciplinary perspectives (see also Chapter 7, this book).

In their book, Felicity Callard and Des Fitzgerald (Extract 3.1) give an account of how interdisciplinary research across the neurosciences, social sciences and humanities has been the authors' most compelling research experience. Their writing draws on a detailed reflection of the implications of boundary crossing and integration between the neurosciences and the social sciences. The questions and insights at the end of each chapter are a useful resource for those seeking for answers in collaborative settings. We have selected an extract from one of their chapters arguing that, rather than aiming for reciprocity in research collaborations with natural scientists, social scientists should acknowledge and accept the power differentials that inevitably exist in such situations.

Matthias Bergmann and co-authors (Extract 3.2) provide a systematic compilation of transdisciplinary case studies that illustrates methods and tools for knowledge integration that can be used to approach complex problems. During the research process, new knowledge is produced by integrating different problem perceptions and knowledge bases from sciences and societal practice; the aim is to contribute to both societal and scientific progress. Bergmann et al's book supports scholars in the conceptualisation and execution of transdisciplinary research projects, and is also of great relevance for teaching. We have extracted their discussion on the dimensions and types of integration necessary for collaborative research.

Daniel Lang et al (Extract 3.3) outline a set of principles for, and challenges of, transdisciplinary research. Considering case studies from Europe, North America, South America, Africa and Asia, the authors synthesise this material and use it to elaborate a conceptual model of an ideal transdisciplinary research process, outlined in Extract 3.3. Over the years, this model has proved to be useful in different settings, and has opened new discussions on how to design a transdisciplinary sustainability research project.

Our final extract comes from an article by Chris Rust (Extract 3.4), which contributes to the discussion on the role(s) arts play in interdisciplinary research, by questioning how research in the creative disciplines might contribute to knowledge and understanding. In this extract, Rust states, 'Research and practice in these fields may deal with matter that changes meaning with time or context, especially in art, where audiences may be expected to complete the meaning of creative works for themselves'. He illustrates, using different examples, how incomplete or tacit contributions to inquiry can be a valuable part of the enterprise of creating knowledge. Attribution of such contributions is difficult, however, and can potentially lead to a lack of acknowledgement of the work of researchers from practice-based disciplines, thereby hampering integration.

EXTRACT 3.1

Callard, F. and Fitzgerald, D. (2015) 'Against Reciprocity: Dynamics of Power in Interdisciplinary Spaces', in *Rethinking Interdisciplinarity Across the Social Sciences and Neurosciences*, London: Palgrave Macmillan UK, Chapter 6.

Fantasies of power

What is little mentioned across these bureaucratic and public-facing formulations is how interdisciplinary collaborations are structured through very different adjudications of epistemic authority (but see the contribution of Ulrike Felt in Mayer, Konig, and Nowotny 2013, 57) – or that, within and beyond particular projects, different actors have access to very different pots of money; that their practices of writing are differently legible, that they are differentially supported by governments and private foundations; that their views carry differential weights within policy communities, and so on.

Things look a little different if one turns to the scholarly literature. In an analysis of an emergent interdisciplinary field crossing economics and geography, for example, geographer Erica Schoenberger worries about how historically nuanced concepts and methods that are central to her own discipline are disregarded in a 'disciplinary imperialism': the new endeavour 'exclude[s] social conflict and power relations, which are both mathematically and ideologically inconvenient ... geography has been excised from an activity that nonetheless calls itself the new economic geography' (Schoenberger 2001, 378). And it is forms of 'social power' – which is to say, in this case, powerful institutions (Harvard and MIT) and disciplines (economics) – that make the difference. Sharachchandra Lélé and Richard Norgaard, drawing on their work in interdisciplinary climate-change projects, put the situation even more starkly: 'there are', they point out, 'significant differences in the manner in which society treats the social and natural sciences … Naturally, the social sciences are seen as irrelevant, boring, and nonrigorous' They go on:

> the social scientists, because they purportedly were not good enough to get into the 'science stream', are often in awe of the natural sciences. The belief of the superiority of the natural scientists is so deep-rooted that whenever social problems have the slightest technical dimension, politicians have traditionally called on only technicians – the natural scientists – to help solve them. (Lélé and Norgaard 2005, 971-972)

Not only do social scientists get side-lined in such projects, on Lélé and Norgaard's account, but natural scientists are asked to fill the breach: 'Charged with providing policy recommendations, [natural scientists] have to make judgments about how society works. They do not have adequate training to do this, but they are perhaps emboldened to do so by their position and are likely to adopt simplistic models of social dynamics' (Lélé and Norgaard 2005, 970).

In one empirical analysis of an interdisciplinary sustainability project, Susan Gardner shows how attention to status and hierarchy ran through the project – with her social science interlocutors giving their answers through 'nervous laughter' versus what Gradner describes as the more self-confident responses of the natural scientists. 'I have some difficulty seeing some of the value of some of the social sciences', one of the natural scientists tells Gardner: 'I have sat through a number of lectures that some of the social science people have given and walked away wondering what the point was' (Gardner 2012, 248). Another tells Gardener that the social scientists on the project are pretty 'squishy' – '"Where were the hypotheses? How much data were gathered? Where are the statistics?"' (Gardner 2012, 248).

To be sure, this is a one-sided account: even if it is not obvious in the literature we have presented above, we have no difficulty in believing (indeed, are assured by our scientist collaborators that it is the case) that collaborating life scientists, too, experience such asymmetries. This occurs both within individual collaborations (one can well imagine the humanist's equivalent of 'Where were the hypotheses?') and beyond them: to collaborate with people who might be a bit 'squishy', after all, is unlikely to represent much a grab for the garlands and esteem of your scientific peers. But rather than cataloguing this back and forth, what we want to do here is undercut the fantasy that might follow on from both sets of complaints. This is the fantasy of power confronted. For Lélé and Norgaard, for example, the key to high-quality interdisciplinary practice is researchers throwing into question the barriers that generate these biases: 'contrary to their disciplinary training', they argue,

> participants [in interdisciplinary projects] need to be self-reflective about the value judgments embedded in their choice of variables and models, willing to give respect

to and also learn more about the 'other', and able to work with new models and taxonomies used by others. (Lélé and Norgaard 2005, 975)

For Schoenberger, geography should revert to its own internal interdisciplinarity (here geography is in a relatively unusual position, we would say), or it could go looking for more interesting economists to talk to (Schoenberger 2001, 379–380). For Bronislaw Szerszynski and Maialen Galarraga, who are especially interested in the case of geo-engineering, the task of social scientists in these situations is to 'expose assumptions, bring out the multiplicity and incommensurability of different views and ontologies, and keep problem definitions open' – thus producing 'greater diversity and reflexivity in how different disciplines and approaches are brought together' (Szerszynski and Galarraga 2013, 2818, 2822). For Dena MacMynowski, clear differentiation is the key to good synthesis: 'in order for interdisciplinary research to proceed more transparently in terms of the recombination of ideas and making the power associated with knowledge claims explicit, interdisciplinary environmental research needs to consciously embark on a process of differentiation and clarification before, or while, moving toward synthesis' (MacMynowski 2007).

From many years thinking about and living in interdisciplinary spaces, we have learnt that this account of power confronted – in which differentials are acknowledged and addressed, and even occasionally overcome; or in which, at least, a model of interdisciplinarity less riven by relations of power is thought to be imaginable – is no less fantastical than the dream of mutuality with which we began. You can have all the frank conversations in the world with collaborators about the conditions under which your exchanges are taking place; you can agree on a clear distribution of resources and labour throughout the collaboration; you can put in agreed strategies to ensure, as far as possible, that this will work; you can remain as open, and transparent, and clear, and dialogic as possible; but the reality is that financial and epistemic power is not distributed equally within the collaboration. That irrevocable state of affairs carries many sequelae in its trail. And everyone in the collaboration, in her heart of hearts, knows this.

After reciprocity

What is it, then, that is desired in subjection? Is it a simple love of the shackles, or is there a more complex scenario at work? (Butler 1997, 27)

What would happen if we gave up on mutuality altogether? What if we abandoned the ideal of 'reciprocity'– not because it isn't realistic, but because it may not actually be desirable? Lately, and instead of adding our own analysis of how 'exchanges' might take place in interdisciplinary spaces, we have begun to wonder if new conceptualizations of interdisciplinarity would come into view if this term were not subject to the velvet stranglehold of mutuality. All of us who have collaborated, after all, have experience of doing so without any expectation either that everyone is on the same footing, or that meaningful forms of process, dialogue, and so on, will actually produce parity. So what if we just accepted, in the most basic sense, that if you're a humanities or social science person in an interdisciplinary neuroscientific project, then there is some likelihood that you're going to have to meet your interlocutors more than halfway, that the emotional labour of holding things together might well fall substantially to you, and that there's a good chance that you're going to have to take responsibility for much of the project 'housework'? What if, as a natural science person, you made peace with the fact that your collaborators will assume you have never thought through your field's problematic use of particular terms, that they will sometimes treat you as a kind of veil of rigour, granting them access to various pots of money, and then expect you, as one of the few scientists who can be bothered to attend 'interdisciplinary' meetings, to consistently represent all that is 'wrong' with your discipline?

[…]

Notes & Queries: 6

Q: In terms of inequalities, asymmetries, and lack of parity, where would you say the particular pressure points frequently lie? What should I be mindful of, if and when I enter interdisciplinary spaces?

A: […] there is a host of (usually) individually minor things that we (singly and together) have learnt to attend to, across a wide number of interdisciplinary collaborations – but which, systemized, and understood as an accumulation, may help to bring the power dynamics of a particular collaboration into understanding. […] An inexhaustive list of things to look out for would certainly include some of the following:

- Interdisciplinary research projects are made possible by a vital, and yet often undervalued and invisible, labour of 'housework' – ranging from paperwork, to dealing with university and research administrations, to ensuring that everyone in the collaboration is okay, and so on. Who does this work, and what value is assigned to it? How – at all – is it compensated? Who can get away with a small contribution to the housework, and who cannot?
- How – through which mechanisms, and drawing in which voices – is intellectual labour parcelled out in the project? Who is included in, or allowed to partake of, conceptual and 'philosophical' discussion – and who is expected to simply collect data?
- Who writes and structures an interdisciplinary grant? Who brings the team together? Who does the tedious labour of working out everyone's costs? Who is responsible (or makes themselves responsible) for getting it in on time?
- What kinds of adjudications are made about what is and isn't an important part of the project's process – and through which mechanisms are those adjudications made explicit or not? For example, is spending time reflecting on how one organizes a workshop as valued as the substantive labour of the workshop itself?
- Who does public engagement work – and how, and why? Sometimes public engagement can be seen as low status work; but sometimes, too, it can be a moment for the assertion of seniority. How is this kind of work positioned and distributed in a given collaboration?

Which kinds of audiences are deemed worthy of cultivating? How – and where – does this cultivation take place?

- Thinking through appropriate 'outlets' for publication can reveal all sorts of dynamics. On the one hand, the logic of impact factors and h-indices often determine where (and by whom) labour is placed in writing; one the other, the general orientation of a journal may make implicit decisions about whose contribution is and is not going to be valued (e.g. how does neuroscientific work get positioned in a decision to submit an interdisciplinary paper to a cultural theory journal?). How – and by whom – do such decisions get made within a collaboration?

- Interdisciplinary, co-authored work of course (usually) needs to find some mid-range voice in which to articulate itself. But dynamics of power can sometimes structure both the relative claim that different disciplinary registers have on that middle range, as well as the pernicious maintenance of the fiction of a 'transparent language' (in other words, that one's one language represents the neutral position). How are registers of appropriate voice generated within the project? What do those voices sound like – and what legacies do they drawn upon?

- Sometimes expertise is considered fungible, and sometimes it is not; sometimes it is regarded as easily understandable, and some it isn't; some expertise is seen not as expertise but merely as 'common sense'. Focusing on such positioning and understandings of different expertise can often be a salutary experience in interdisciplinary work.

- When papers are co-authored, how and when do particular conventions from different disciplines get pulled in? Does the team follow the multi-authored conventions of the natural sciences (and the stratification of contribution that it implies), or does it follow the lone muse humanities model (and the occlusion of other voices that thus becomes inescapable)? If artists are part of the collaborating team, has anyone worked to rethink practices of creative production (in relation to curatorship)?

EXTRACT 3.2

Bergmann, M., Jahn, T., Knobloch, T., Krohn, W., Pohl, C. and Schramm, E. (2010) 'The Integrative Approach in Transdisciplinary Research', in *Methods for Transdisciplinary Research: A Primer for Practice*, Frankfurt/Main: Campus Verlag, Chapter 1.

Dimensions and types of integration

A study of research reports and publications on transdisciplinary research reveals a number of strategies and methodologies supportive of an integrative transdisciplinary research process. These include theoretical frame-works, interdisciplinary hypothesis formulation (including practical input from societal actors), model building, integrative assessment procedures, the development of new cross-disciplinary methods and the building of a process organization supportive of integration. In terms of integration, much has already been developed and applied. [...] Scientists tend to underestimate the work required of them to ensure the integration needed in transdisciplinary research processes, with this integration normally being understood primarily, or even exclusively, as knowledge integration. In fact, however, there are several interrelated dimensions in which integration work must be accomplished:

- A *communicative* dimension: This is the (differentiating and) linking of different linguistic expressions and communicative practices, with the aim of developing something like a common discursive practice in which mutual understanding and communication is possible, as well as clarifying common terms and constructing new ones.
- A *social and organizational* dimension: Here it is a matter of differentiating and correlating the participating researchers' different interests and activities, as well as of the sub-projects or organizational units. This dimension also includes the conscious leadership of (not only scientific) teams, mutual understanding and the willingness to learn.
- A *cognitive-epistemic* dimension: Here it is a question of the differentiation and linkage of expert/disciplinary knowledge bases, as well of scientific and practical real-world knowledge. More concretely: it is a matter of understanding the methods and terms of other disciplines; clarifying the limits of one's own knowledge; and developing methods and building theories together.

Cutting across these integration *dimensions* one can identify *types* of integration in transdisciplinary research. While with the concept "dimensions" we describe, in relatively general terms, specific challenges facing integration, with the concept "type" we can already begin to indicate the concrete occasions and typical integration constellations that limit and order the need for integration.

[...]

Symmetric integration

This type of integration is found wherever specialist disciplines supply different parts of the whole of a body of knowledge needed, whether from their existing stocks of knowledge or by producing new knowledge specifically for the purpose. If we take, for example, the model project, Fischnetz (see III.D) we see that researchers from environmental chemistry, ecotoxicology, population biology, limnology, hydrology, and others were all involved. Their skills and knowledge complement one another like the pieces of a puzzle, albeit with overlaps and gaps. The overlaps pose communication problems, because here the phenomena as grasped by different research fields are interpreted in partially different languages. The gaps, on the other hand, indicate where research is needed. The need for integrative work is foreshadowed by the very fact that individual aspects of the overall problem and their handling have been allocated to specialist disciplines at the beginning of the research process. A key task, then, facing integrative research is to decompose the overall problem into individual aspects that are manageable by the specialist disciplines and, at the same time, to ensure a consistent melding of the partial results provided by each discipline.

Integration of social and natural sciences

If the natural and social sciences play a nearly equally important role in a project, this almost always causes serious integration problems. Contrary to an often held opinion, this is not because of the famous "different cultures," with their mutual prejudices. The sharpest points of contention lie between closely related disciplines or scientific fields, not between those widely separated. The crucial point is not "cultural" differences but the heterogeneity of knowledge bases, something which blocks the kind of working together characteristic of symmetric integration. Almost all the key terms used in social-ecological projects have both an objective sense, embedded in the natural sciences, and a subjective sense, embedded in the social sciences. While natural-sciences-related aspects can be linked to actual, detectable causal relationships such as the increase or decrease of grasslands, floods and the spread of disease, the social-science-related aspects are linked to things more difficult to detect and measure, such as perceptions, attitudes, expectations, habits and beliefs. The *concept of risk* is the best known example of this duality. Data collected by insurers on the volume of insurance claims and the probabilities of risk cannot be directly projected onto the data collected by social science on the perception and fear of risks, and certainly not onto the study of risk taking. The collection of both types of data is justified because each is in its own way informative. However, it is difficult to articulate a coherent, quasi additively integrated system of statements, because categories are defined at different levels. They do not complement one another, as do, for example, categories found in the hydrological and geological sciences.

[...]

Integration of formal and empirical sciences

A different kind of integration problem arises when the skills and knowledge of formal and empirical sciences must be coordinated. This problem arises already in the disciplinary context and is one reason why work in disciplines and sub-disciplines may assume interdisciplinary features. Specialists in computer science, statistics or instrument engineering are regularly found working in physics, biology or sociology, and their activities must be negotiated and coordinated. In particular, there must be a continuous negotiation between theoretical goals and methodological possibilities. However, in disciplinary contexts, integration often takes the form of well rehearsed symbioses (for examples, gene sequencing in biology, the Monte Carlo method in physics, structure analysis in the sociology or Bayesian probability in analysis of error). Transdisciplinary research, in contrast, must not only search for appropriate methodologies but often enough create these itself. A prime example of this need for total methodological innovation (with its opportunity for new methodological insights) in the context of integrating the individual perspectives of formal and empirical sciences can be found in model project 9 (see III.J) which deals with communication research. There, the formal sciences include computer science and mechatronics, with the empirical sciences represented by linguistics and psychology. Similar situations are found in many transdisciplinary fields, for example, in the field of energy production (power plant control, renewable energies) or when dealing with genetically modified plants.

Theoretical and conceptual integration

From the perspective of the philosophy of science the integration processes that are the most challenging and difficult to implement are those in which a new conceptual foundation must be found, one which goes beyond existing knowledge corpora or existing disciplines. Such projects raise fundamental questions already during the scientific reconstruction of real-world problems into scientific research questions. If these questions are ignored at the beginning they will become even more pressing later. A particularly urgent current example of this challenge is climate change. Many disciplines have contributed explanations of the sub-processes involved in climate change, but an integrated understanding of the overall process is still far from our grasp. Among the projects presented in this book, relevant examples include the search for a "multi-modal interactive communication concept" in the project "Situated human-machine communication" (see III.J); the development of a comprehensive understanding of social mobility as a primary societal relation to nature (see III.B); and the concept of "urban design" in the project "Synoikos" (see III.H). Since the theoretical resolution of such fundamental questions is not always possible within a normally rather short-term research project, a pragmatic selection of medium range goals is necessary.

EXTRACT 3.3

Lang, D.J., Wiek, A., Bergmann, M., Stauffacher, M., Martens, P., Moll, P., Swilling, M. and Thomas, C. (2012) 'Transdisciplinary research in sustainability science: Practice, principles, and challenges', *Sustainability Science*, 7: 25–43.

Concept of an ideal–typical transdisciplinary research process

Key components of an ideal–typical transdisciplinary process are here presented in order to position the derived principles as accurately as possible *within* the actual research practice. We rely in this article on a slightly adapted version of an ideal–typical conceptual model (Jahn 2008), which has many similarities to other model presented in the literature (e.g., Scholz et al. 2006; Pohl and Hirsch Hadorn 2007; Wiek 2009; Carew and Wickson 2010; Krütli et al. 2010b; Stokols et al. 2010; Talwar et al. 2011). According to this model (Fig. 1), transdisciplinary research in general and in sustainability science in particular is an "interface practice": first, it initiates from societally relevant problems that imply and trigger scientific research questions; second, it relies on mutual and joint learning processes between science and society embedded in societal and scientific discourses (Siebenhüner 2004). In so doing, transdisciplinary research integrates two pathways to address "real world problems": one pathway is committed to the exploration of new options for solving societal problems (the path of problem solution, the left "arm" in Fig. 1); the other pathway is committed to the development of interdisciplinary approaches, methods, and general insights related to the problem field (the path of scientific innovation, the right "arm" in Fig. 1), which are crucial for the practical path (cf. Bergmann et al. 2010).

In the ideal–typical conceptual model presented in Fig. 1, a transdisciplinary research process is conceptualized as a sequence of three phases, including: collaboratively framing the problem and building a collaborative research team (Phase A); co-producing solution-oriented and transferable knowledge through collaborative research (Phase B); and (re-)integrating and applying the produced knowledge in both scientific and societal practice (Phase C). Thereby, a main purpose of Phase A is to integrate "the pathway of problem solution" and the "pathway of scientific innovation" to allow for collaborative research in Phase B ("integrative research pathway"), resulting in transferable knowledge that can be (re-)integrated into the societal and scientific practice in Phase C. Though the model might indicate a rather linear process, individual phases and the overall sequence often have to be performed in an iterative or recursive cycle, also highlighting the need for reflectivity in transdisciplinarity (see, e.g., Spangenberg 2011). In this article, we slightly adapt the original model by: (a) changing the terminology to match the international discourse in sustainability science and related fields and (b) underlining the need for a deliberate design of the collaboration between actors from academia or other research institutions and actors from practice.

Phase A: Collaborative problem framing and building a collaborative research team

This phase orients, frames, and enables the core research process. It consists of several activities: identification and description of the real-world problem; setting of an agreed-upon research object, including the joint formulation of research objectives and specific research- as well as societally-relevant questions; the design of a conceptual and methodological framework for knowledge integration; and the building of a collaborative research team. Essential in this phase is that the real-world problem is translated into a boundary object (see, e.g., Clark et al. 2011) that is both researchable and allows for the re-integration of the insights into societal implementation as well as the scientific body of knowledge.

Phase B: Co-creation of solution-oriented and transferable knowledge through collaborative research

This phase is the actual *doing* of the research. In this phase, a set of integrative (scientific) methods is adopted, further developed, and applied to facilitate the differentiation and integration of the different bodies of knowledge coming together in the process. Concomitantly, a collaborative research design allows for goal-oriented collaboration among different disciplines, as well as between researchers and actors from outside academia, in a functional and dynamic way. For each step of the research process, it needs to be defined who contributes what, supported by which means and to what end (Krütli et al. 2010a, b). Thereby, it is important to consider different levels of stakeholder involvement in the research process (Wiek 2007; Stauffacher et al. 2008; see the (red) zigzag line in the center of Fig. 1).[1]

Fig. 1 Conceptual model of an ideal-typical transdisciplinary research process (adapted from: Bergmann et al. 2005; Jahn 2008; Keil 2009; Bunders et al. 2010; there are several models which outline transdisciplinary research process in a similar way: e.g., Scholz et al. 2006; Pohl and Hirsch Hadorn 2007; Wiek 2009; Carew and Wickson 2010; Krütli et al. 2010b; Stokols et al. 2010; Talwar et al. 2011)

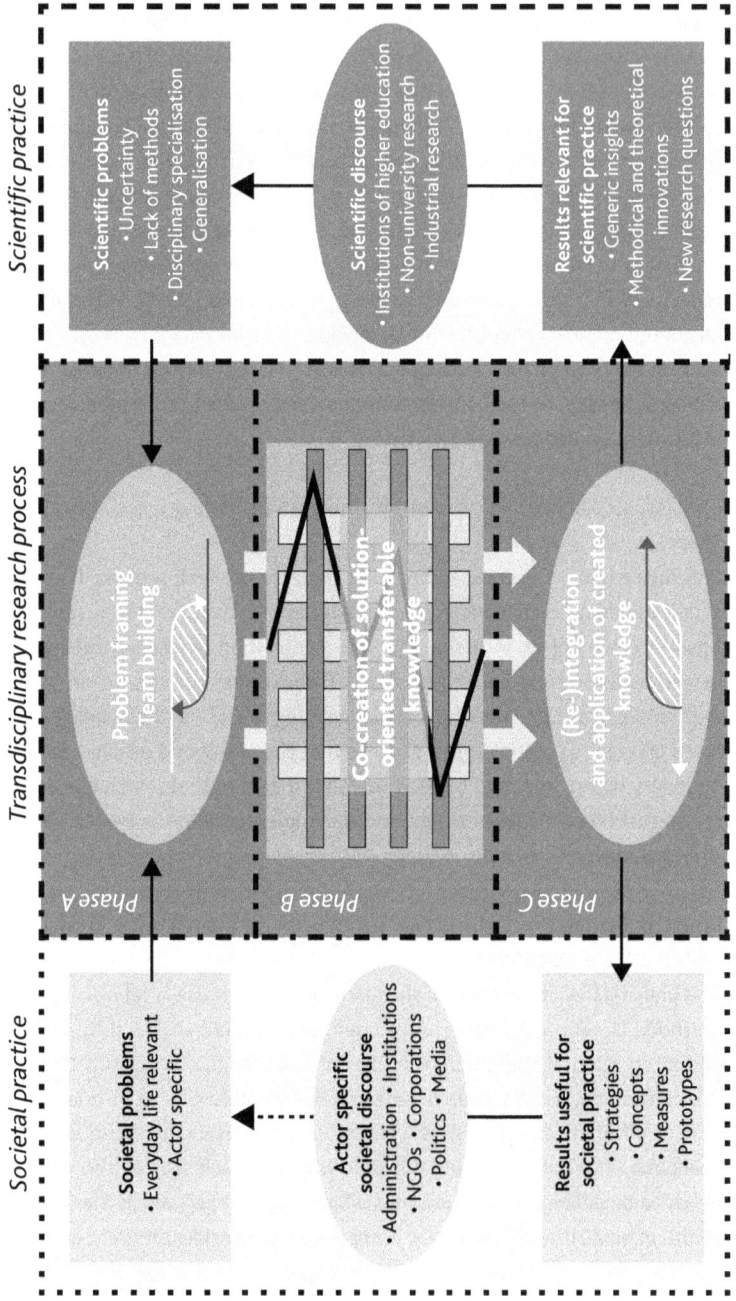

Scientific practice

Scientific problems
• Uncertainty
• Lack of methods
• Disciplinary specialisation
• Generalisation

Scientific discourse
• Institutions of higher education
• Non-university research
• Industrial research

Results relevant for scientific practice
• Generic insights
• Methodical and theoretical innovations
• New research questions

Transdisciplinary research process

Phase A
Problem framing Team building

Phase B
Co-creation of solution-oriented transferable knowledge

Phase C
(Re-)integration and application of created knowledge

Societal practice

Societal problems
• Everyday life relevant
• Actor specific

Actor specific societal discourse
• Administration • Institutions
• NGOs • Corporations
• Politics • Media

Results useful for societal practice
• Strategies
• Concepts
• Measures
• Prototypes

Phase C: (Re-)integrating and applying the co-created knowledge
This phase is the process of using, applying, and implementing the research results. As different perspectives, world views, values, and types of knowledge are integrated over the course of the entire transdisciplinary research process, this phase is not a classical form of knowledge transfer from science to practice (van Kerkhoff and Lebel 2006; Talwar et al. 2011). It is, instead, a (re-)integration of the results into: (a) the societal practice (e.g., implementation of the evidence-based strategies and action programs generated during the research) and (b) the scientific practice (e.g., comparison, generalization, and incorporation of results into the scientific body of literature). Apart from the tangible products (e.g., strategies), the transdisciplinary process, if designed accordingly, might also lead to less tangible but equally important outcomes, such as enhanced decision-making capacity of the practice actors involved (Wiek et al. 2006; Walter et al. 2007). Likewise, a transdisciplinary project can trigger an intense learning process. It can empower and motivate stakeholders to contribute more actively to the implementation or related decision processes. Especially in the field of sustainability science, the centralized steering idea has to be questioned and, in many cases, be replaced by the metaphor of an ongoing learning process (see, e.g., Laws et al. 2004).

Design principles for transdisciplinary research in sustainability science

[...]

The set of design principles is structured into the three phases of a transdisciplinary research process introduced in the previous section (Table 1). We have formulated those principles close to the actual research practice and as tasks that can be assigned to specific actors (researchers, stakeholder, facilitators, etc.) along the three phases of the research process. There is no fixed rule on who should take the lead with regards to which task; yet, it is important to assign responsibilities right from the beginning of the project (see the principles of *team building* and *assignment of appropriate roles*). To make the design principles as well as the phases more tangible, a transdisciplinary "model project" is presented along the principles in Table 2.

Phase A: Design principles for collaborative problem framing and building a collaborative research team
• *Build a collaborative research team.* Identify scientists from relevant disciplines/ scientific fields and "real-world actors" who have experience, expertise, or any other relevant "stake" in the problem constellation pre-identified for the research project (Pohl and Hirsch Hadorn 2007). Apply transparent criteria and justifications for who should and who should not be included in the research project and why. Often, this is a recursive process when expanding the team with additional experts or real-world actors representing specific interests, expertise, or experiences *after* the initial problem description. Facilitate explicit team-building processes (*Selecting* team members and *building* a collaborative team are two different steps in the

Table 1. Design principles for transdisciplinary research in sustainability science and related guiding questions

Design principle	Guiding question
Phase A	
Build a collaborative research team	Does (did/will) the project team include all relevant expertise, experience, and other relevant "stakes" needed to tackle the sustainability problem in a way that provides solution options and contributes to the related scientific body of knowledge?
Create joint understanding and definition of the sustainability problem to be addressed	Does the project team reach a common understanding of the sustainability problem to be addressed and does the team accept a joint definition of the problem?
Collaboratively define the boundary/research object, research objectives as well as specific research questions, and success criteria	Is a common research object or guiding question, with subsequent specified research object and questions, formulated, and does the partners agree on common success criteria?
Design a methodological framework for collaborative knowledge production and integration	Does the project team agree upon a jointly developed methodological framework that defines how the research target will be pursued in Phase B and what transdisciplinary settings will be employed? Does the framework adequately account for both the collaboration among the scientific fields and with the practice partners?
Phase B	
Assign and support appropriate roles for practitioners and researchers	Are the tasks and roles of the actors from science and practice involved in the research process clearly defined?
Apply and adjust integrative research methods and transdisciplinary settings for knowledge generation and integration	Does the research team employ or develop methods suitable to generate solution options for the problem addressed? Does the team employ or develop suitable settings for inter- and transdisciplinary cooperation and knowledge integration?
Phase C	
Realize two-dimensional integration	Are the project results implemented to resolve or mitigate the problem addressed? Are the results integrated into the existing scientific body of knowledge for transfer and scaling-up efforts?
Generate targeted products for both parties	Does the research team provide practice partners and scientists with products, publications, services, etc. in an appropriate form and language?
Evaluate scientific and societal impact	Are the goals being achieved? What additional (unanticipated) positive effects are being accomplished?

(continued)

Table 1   Design principles for transdisciplinary research in sustainability science and related guiding questions (continued)

Design principle	Guiding question
General Design Principles (cutting across the three phases)	
Facilitate continuous formative evaluation	Is a formative evaluation being conducted involving relevant experts related to the topical field and transdisciplinary research (throughout the project)?
Mitigate conflict constellations	Does the researchers/practitioners prepare for/anticipate conflict at the outset, and are procedures/processes being adopted for managing conflict as and when it arises?
Enhance capabilities for and interest in participation	Is adequate attention being paid to the (material and intellectual) capabilities that are required for effective and sustained participation in the project over time?

The precise formulation of the design/evaluative guiding questions depends on the specific type of evaluation, e.g., ex-ante assessment, formative evaluation during the research process, or ex-post evaluation (internal or external)

overall process.). Furthermore, it is crucial to establish an organizational structure in which responsibilities, competencies, and decision rules are clearly defined. In many cases, a good strategy is to establish balanced structures between researchers and practice actors on all organizational levels including a joint leadership (see, e.g., Scholz et al. 2006). Make sure to contract in advance professional facilitators who can support the team at critical stages of the research process. A key aspect of building a collaborative team is to develop a "common language" among all team members. This is a joint effort that builds capacity and prevents misunderstandings and roadblocks for collaboration at later stages of the research process. This effort cuts across the subsequent activities (problem definition, etc.) and continues into Phase B. Key components in this respect are, first, to commonly define those terms that play a central role in the problem field and/or are used differently in collaborating disciplines and, second, to build a joint understanding of key concepts relevant in the research process (Stokols et al. 2010).

- *Create joint understanding and definition of the sustainability problem to be addressed.* Define the sustainability problem as a societally relevant problem that implies and triggers scientific research questions. Justify that this is, in fact, a sustainability problem and not just any kind of complex problem (Siebenhüner 2004; Wiek et al. 2012). Make sure all team members (scientists and practitioners) are involved in the problem definition. Facilitate the process in a way that integrates and balances "contradicting normative scientific and political claims of importance and relevance" (Bergmann and Jahn 2008, p. 92). This sub-principle ensures that any subsequent research task departs from this common reference point and, thus, contributes to the overarching project goal.

- *Collaboratively define the boundary/research object, research objectives as well as specific research questions and success criteria.* Collaboratively formulate the overall object and objective(s) of the research process in order to be able to track progress and realign research activities during the research process (Defila et al. 2006; Blackstock et al. 2007). The boundary/research object can be formulated in a guiding question (Scholz et al. 2006) and often needs to be further specified in a set of sub-questions. The definition of the research objective(s) requires explicitly accounting for the different interests of scientists and practitioners collaborating in the project (Wiek 2007). While synergistic, both parties pursue ultimately *different* objectives (extending the body of scientific knowledge vs. solving/transforming the real-world problem), and it is advantageous for the process to make these differences transparent. Still, especially the roles of scientists will become multiplied and potential role conflicts need to be reflected upon. Subsequently, the boundary/research object and the objectives needs to be further specified into operationalized research questions, which is a crucial step for developing an integration model, and facilitates designing the methodological framework. Outline the success criteria, which will be used to evaluate whether the objective(s) was/were met or not.
- *Design a methodological framework for collaborative knowledge production and integration.* Agree on the set of methods and transdisciplinary settings to be applied in Phase B and develop a concept for integrating the research results throughout the project (see, e.g., Scholz et al. 2006; Wiek and Walter 2009; Talwar et al. 2011). Existing methodological compilations for transdisciplinary research should be consulted (see, e.g., Scholz and Tietje 2002; Weaver and Rotmans 2006; Bergmann et al. 2010). The latter concept should also employ evidence-based templates for collaboration, such as the functional–dynamic model of participation proposed by Stauffacher et al. (2008). Such a framework allows a structured collaboration and synthesis across all team members and project phases. The framework might have to be adjusted during the project, but it provides a common orientation for all team members from the beginning (Defila et al. 2006; Scholz et al. 2006).

Phase B: Design principles for co-creation of solution-oriented and transferable knowledge through collaborative research

- *Assign and support appropriate roles for practitioners and researchers.* Assign in each research effort appropriate roles and responsibilities for scientists and practitioners in a transparent process, accounting for inertia, reluctance, and structural obstacles (Maasen and Lieven 2006; Wiek 2007). Base the assignments on the overall framework outlined in Phase A and make sure that they comply with the predefined organizational structure of the project. For scientists, balancing societal relevance with scientific rigor becomes a key challenge and asks for particular attention. Ensure facilitation that allows compliance with the assigned roles and responsibilities as well as attaining to the aspired levels of participation (van Kerkhoff and Lebel 2006; Wiek 2007). Furthermore, leadership related to cognitive (providing a means to integrate the different epistemics of the actors involved), structural (addressing the needs for coordination and information exchange), and procedural (resolving conflicts during the process) tasks facilitates successful transdisciplinary processes (Gray 2008).

• *Apply and adjust integrative research methods and transdisciplinary settings for knowledge generation and integration.* According to the methodological framework developed in Phase A, make use of and further develop appropriate methods for transdisciplinary sustainability research. Use tools to support teamwork and collaboration such as the advocate principle, the tandem principle 2, or the mentors' principle (see Table 2) (Bergmann et al. 2010). Such instruments provide the research team with valuable support for inter- and transdisciplinary quality control and help to make research results better accessible for practice partners. The team might also utilize their collaborative potential and further develop existing or develop novel methods for transdisciplinary knowledge production and integration.

Phase C: Design principles for (re-)integrating and applying the created knowledge

• *Realize two-dimensional (re-)integration.* Review and revise the outcomes generated in Phase B from both perspectives separately, i.e., the societal and the scientific practice. Likewise, the mutuality of the learning process becomes visible. It is important to employ different criteria for revision and rendering, as both perspectives adhere to quality criteria such as scientific credibility or practical applicability (saliency) differently (Wiek 2007; Jahn 2008).

• *Generate targeted "products" for both parties.* Provide the scientific actors and the practice partners with appropriate products (cf. Defila et al. 2006) that present and "translate" the results of the project in a way that the actors can make use of—as a contribution to real-world problem-solving/transformation or to scientific progress/innovation (Pohl and Hirsch Hadorn 2007).

• *Evaluate societal and scientific impact.* Evaluate the project at different stages after completion of the project to demonstrate impact and generate lessons learned for future project design (Walter et al. 2007). For both scientific as well as societal impacts, an important reference point is the success factors defined in Phase A, which might have been adapted in the course of the project.

General design principles cutting across the three phases

• *Facilitate continuous formative evaluation.* Formative evaluation throughout the transdisciplinary sustainability research project by an extended peer group (comprising experts from science and practice) allows reviewing progress and reshaping the subsequent project steps and phases if necessary (Bergmann et al. 2005; Walter et al. 2007; Regeer et al. 2009).

• *Mitigate conflict constellations.* Transdisciplinary research is characterized by continuous interaction between scientists from different disciplines and different practice actors. The context that existed at the outset of the process can rapidly shift as new actors become involved, actors change roles or attitudes, new insights are being revealed, and so forth. In order to prevent conflicts, reflexive meetings, open discussion forums, explicit and mediated negotiations as well as adapted agreements should accompany the transdisciplinary research process over the entire course of the project (van den Hove 2006; Wiek 2007). This means that the learning process inherent in a transdisciplinary project needs to be carefully designed and followed.

• *Enhance capabilities for and interest in participation.* It cannot be assumed that all actors have the capacity or continuous interest to participate in a given transdisciplinary research project that might continue over several years. Some actors might underestimate the time and energy necessary to participate in a meaningful way, while others might not have the means to become involved from the outset. For continuous stakeholder participation it is inter alia important: to select locations that are easily accessible for stakeholders; to schedule meeting times that allow maximum participation; to facilitate discussions in several languages (as necessary); to involve stakeholders through a high level of interactivity that allow participants not only to articulate their perspectives, but also to engage in meaningful discussions, deliberations, and negotiations; and to incorporate visual products and media, for instance, by visual designers during the activities in order to allow for meaningful interactions across different languages, levels of literacy, and educational backgrounds (Stokols et al. 2010).

The purpose of the design principles formulated above is to practically guide transdisciplinary research processes and facilitate an effective and efficient research process for all actors involved. The principles present ideal–typical guidelines rather than instructions that can be applied in any given context.

EXTRACT 3.4

Rust, C. (2007) 'Unstated contributions: How artistic inquiry can inform interdisciplinary research', *International Journal of Design*, 1(3): 69–76.

The Problem of Contribution

[...]

In his analysis, referred to above, of the contribution that designers and artists can make to research in the natural sciences, the author (Rust, 2004) building on Polanyi's work on the theory of tacit knowledge, described how the imaginative creation of "new worlds" can help scientists to identify and commit themselves to new hypotheses. That analysis provided case examples of designers creating artefacts that, when put to use in some way, revealed new ideas and research opportunities to different individuals, depending on their particular experience and concerns. Clearly, it was necessary for the scientist to "complete the meaning" for themselves. The designer could not achieve that and their work was done, for the time being, once the artefact was deployed. However, the designer quite clearly owned both the initiative to create the new world and the knowledge and inquiry that were embedded in it, this often being gained through a difficult and rigorous programme of contextual research and experiment.

Other collaborations in projects in which the author has had some direct involvement or oversight, reveal variations on this theme. Peter Ainsworth, who investigates mediaeval French literature, has formed a partnership with Colin Dunn, a photographer, to capture images of important manuscripts from the 100 Years War between France and England, 1337-1453. These are precious and fragile documents, rich in text, illustration and artistry, which are difficult to study as they are kept in controlled conditions in museums and libraries scattered around Europe. By capturing very high resolution digital images of every page from several of these manuscripts (Figure 3), Dunn and Ainsworth have made it possible to study all of them in one place with little practical difficulty. Ainsworth is able to summon any page to a large screen in his office and zoom in on any part in very fine detail. In this case, it is clear that Peter Ainsworth both owns the analysis of the manuscripts and is responsible for knowing.

To carry out this work Dunn had to use special equipment and techniques developed in earlier projects. His expertise and insight, including experience of working as a calligrapher, was also needed in photographing the manuscripts to ensure that tiny details, some of them 3-dimensional in nature, were revealed so that Ainsworth could do his work of detecting clues to the production and meaning of the books. Dunn also produced software to assist in viewing and organizing the images. Ainsworth can only examine what Dunn is able to show him. Without Dunn's knowledge and insight as a photographer, arguably comparable to if different from Ainsworth's knowledge and insight as a mediaevalist, the research would not happen. Dunn may not own its conclusions but he has a clear stake in its foundations. It might be argued that Colin Dunn makes no more contribution than the lens grinder or camera builder but their work is generic, based on existing knowledge. Dunn has studied the actual material of this research and transformed it in line with his understanding of Ainsworth's aims.

Dunn and Ainsworth are each properly aware of their limits and their need for each other. Dunn has the ability to reveal what Ainsworth wishes to examine, but it is absolutely vital that he does not attempt to say what is significant in the landscape he has uncovered. Only Ainsworth can complete the meaning. What is equally important is that another specialist in history, language, art, technology, society or so forth may be able to discover completely different insights in the same material. Again, Dunn has no way to predict what these will be. His contribution can only be to frame the revelation. Any attempt to go beyond that on his own account is not relevant and could be damaging. It would be easy to say that Dunn's contribution is secondary to Ainsworth's, but that assumes that Ainsworth is making the only significant analysis of the material Dunn has captured. If a great many future scholars were to exploit Dunn's work we may yet conclude that his was the essential contribution.

For a further example, which deals with communication between individuals of different disciplines, recent work on creating valid visual metaphors for the molecular actions of nanotechnology illustrates the importance of tacit transmission. A design group consisting of Jeff Baggott, a filmmaker, and Nick Dulake, an industrial designer with specialist skills in computer modelling, have worked with a group of scientists to create video material (eg. Baggott, Dulake, Jones, & Ryan, 2005) that provides a general audience with an understanding of effects that physicists would normally describe through mathematics. It is not possible to create an intelligible, literal 3-dimensional representation of these molecule-scale events and the metaphors that had been used previously were naive and, if anything, impeded understanding of the science.

Dulake described their process (personal communication with the author, June, 2007), making it clear that the physicist and he lacked any shared formal language to deal with the situation. Instead, he uses the very limited understanding he can glean from the scientists', largely incomprehensible, mathematical descriptions to create a tentative sketch for a possible visual metaphor. From the conversations that ensue, he and Baggott gradually refine and direct his efforts until the physicist is satisfied that the visualisation (e.g. Figure 4) is a valid metaphor of the principles at work. The second stage of the work, led by Baggott, is to sustain that metaphor into a time-based work, a video, which demonstrates the nanotechnology actions and relates them to their human scale effects. Again, the aim is always to ensure that the visual narrative remains true to the physicists' scientific understanding while being meaningful to the audience.

The reason for introducing this account is to show how tacit or visceral communication can be the only way that some knowledge can be transmitted, even in a natural science setting. The physicists must look for reflections of their own understanding in the designers' non-literal representation. The designers must detect the physicists' meaning, despite having no real grasp of their language. Finally, the eventual audience must gain a useful, "true" sense of the physicists' knowledge from the designers' work, even though it contains no true factual information and uses a novel metaphor. The designers have a well-refined expertise in creating non-verbal communication that also relies heavily on tacit insight on their part.

The examples above illustrate different aspects of the problem at hand but they are all partial and inconclusive. The final example, by contrast, is of research undertaken explicitly to engender insights in others. Lucy Lyons (2006) conducts research that aims to reveal new understanding of a particular disease, Fibrodysplasia Ossificans Progressiva (FOP). This is a dramatic physical condition that affects very few people. Lyons has taken the initiative to record a diverse set of human remains and living sufferers through the use of drawings, which she describes as "delineations" (Figure 5). The practice of drawing what she observes in a specific case contrasts with the more generalised conventional medical illustrations produced to describe established knowledge. Lyons has taken the initiative to track down the skeletons and specimens in a variety of locations, including non-medical museums. She has built up a network of interested parties, including pathologists and patients, and developed a methodology that includes working in partnership with a specialist technician who is macerating (dissecting and preserving) a cadaver while she draws it. She bases her work partly on that of the 19th Century pathologist Richard Carswell whose exceptional drawings revealed the physical nature of a disease and insights into the experience of sufferers. Lyons's work so far has been validated partly by confirmation from pathologists that her delineations do reveal new insights into the physiological effects of the disease.

Unlike Colin Dunn, Lyons owns the whole process up to the point where she positions the work to allow the scientists to become involved with it. It can be argued that she "owns" the result, but however much she engages with pathologists and what they know the one thing she cannot and should not do is to predict what they will learn from her work.

Conclusion—Tentative Principles

Grounded in the examples presented above, a set of tentative principles is offered that indicate how unstated or generative contributions might operate, especially in interdisciplinary research. With the exception of the work of Lucy Lyons, no claim is made that any of these examples, as briefly presented in this paper are valid as research in themselves, although in several cases they are drawn from wider research projects that can stand up to more detailed scrutiny. An example is the quality of communication achieved by Baggott and Dulake, which validates their hypothesis that their novel methods of working with scientists will generate valid and useful metaphors for difficult scientific concepts.

It is proposed that there can be valid research whose contribution to knowledge cannot be stated fully or precisely by the researcher. This is particularly relevant to research by creative artists, but it also has implications for interdisciplinary or multidisciplinary research that might result in contributions in different domains and where not all participants can "own" the conclusions unless their partners are prepared to acknowledge the importance of the developmental contributions.

The underlying principle is that some contributions are necessarily generative, providing a point of departure for others in the sense that Dunn provides the environment for Ainsworth's analysis and Lyons provides a means for pathologists to examine a disease. The artist's position that the viewer completes the meaning is compatible with this. It is not necessary for these generative contributions to be specific. As discussed above, the value of Colin Dunn's work lies in providing access to a "landscape" of material and it is important that he makes no strong judgement about what might be significant in that landscape, although he will do his utmost to reveal as much and as many different kinds of data as he can.

The nature of the contribution can vary. For example, the work may draw attention to an issue or engender insights into that issue. In Ainsworth's research some of the fine detail revealed enabled him to identify people involved in the production of the manuscripts and some organisational aspects of their production. It may provide a resource for reflection or analysis as in Lyon's work aimed at pathologists. It may indicate directions or techniques for the disciplines concerned or it may provoke critical reflection on the audience's own situation. These kinds of contribution, or the possibility of them, can be seen in the material discussed above, but this is not intended to be an exhaustive list. The most important feature is that while they should result from an intentional inquiry there is no need for them to be predicted by the researcher except in the broadest terms. Only the audience can determine what is relevant.

The methodology for such research must acknowledge the role of tacit transmission. Completing the meaning of the work is likely to draw on the tacit knowledge of the viewer.

As previously described (Rust, 2004) the process may reveal aspects of that tacit knowledge as an outcome in itself. Similarly, as set out above, the inquiry itself may have to rely on indwelling and tacit acts of translation between, for example, observation and synthesis, or in the social process between researcher and subject or collaborator, as in the work by Dulake and Baggott.

Finally, having acknowledged the need for artists and some others to avoid convergence in their research, we come to the quid pro quo implied by the artists' workshop that identified this issue. Those who wish to be regarded as researchers—as well as being artists or photographers or designers—must "own" their research in several important ways. They must declare the subject of their inquiry and their motivation for investigating it. They must demonstrate that they have a good understanding of the context for the work and what has gone before. They must have both methods and methodology and they must set all these things out in ways that the rest of us can recognise and understand, although we need not be prescriptive about the actual means of doing that.

Beyond that, any researcher would be wise to attend to the consequences of their work. An artist may not predict the results of their contribution, but after the event they have the opportunity to inspect what has happened and own it. Whether they do this by their own efforts or by ensuring that suitable others are doing the work required is less important than that it is done. Simon Bowen's research embodies the useful principle of tacitly processing the events following one "artistic" work into a subsequent work, crystallising some of the potential of the first while opening up new areas of uncertainty. This might provide a more suitable way for the artist to observe and "analyse" the consequences of their work than engaging in a perhaps unnatural act of explicit scientific analysis.

Commentary

Sibylle Studer

Co-creating a research project is challenging. Sometimes it feels like chasing a utopia. You know that co-creation is a time-intensive endeavour, while the actors you want to co-create with have busy time schedules. You want to jointly co-create meaningful results, but you realise how difficult it is even just understanding each other and that a common language is missing. How can learning from others by reading texts help to seize these challenges?

The selected extracts in this chapter may support co-creators as they help to put intentions to co-create into perspective. They show different facets of co-creation, elaborate on the purpose of co-creation, show that it is sometimes also desirable or even necessary not to co-create everything, and offer refreshing views on the roles of the arts, humanities and social sciences in co-creation processes. They provide solid background information to understand the pertinence of rules of thumb that capacity builders may state, such as 'consider co-creation at the very beginning of your project' or 'provide transparency about when, why and with whom you want to co-create'.

I often refer to Matthias Bergmann et al's (Extract 3.2) and Daniel Lang et al's (Extract 3.3) texts in my daily work, for example when I provide capacity building based on the td-net toolbox for co-producing knowledge.[2] Bergmann et al's elaboration on dimensions and types of integration serves to illustrate the need for special instruments and methods that support inter- and transdisciplinary research projects – such as those compiled in the td-net toolbox and elsewhere.[3] Lang et al's ideal-type and principles of process design help to frame and embed the application of specific tools and methods.

The td-net toolbox (Pohl and Wülser, 2019; Studer and Pohl, 2023) contains methods and tools that have the potential to bridge different thought styles (Fleck, 1979). They are meant to be user-friendly since they use mainly everyday language and low-tech equipment. They often consist of structured workshop formats that introduce a heuristic (for example, the 'three types of knowledge' or 'three outcome spaces')[4] to guide the joint elaboration in a transdisciplinary research project. Several methods of the td-net toolbox can be perceived as instruments that help to establish a common ground for ongoing or future co-creation (in iterative cycles),[5] rather than as standalone methods. Tools are needed to make different perspectives explicit, to better understand how others perceive a problem and what others mean by using specific terms and jargon.

Therefore, the td-net toolbox is a real example of the conceptual grey zone discussed by Bergmann et al in Extract 3.2; the grey zone – or spectrum – between instruments that 'although reproducible, can be best described in terms of a more practical and process-oriented approach' and 'reproducible methods that follow a controlled scientific procedure'. Bergmann et al argue that such integration methods and instruments 'help ... to solve the integration tasks found in problem-oriented transdisciplinary research' once the dimensions and types of integration that illustrate the challenges and versatility of integration tasks have been elaborated.

Acknowledging that transdisciplinary research may better rely on combinations of tools and methods (rather than on the rigid application of a single, reproducible, scientific procedure), we often refer to the process described by Lang et al in Extract 3.3. These authors provide an overarching 'ideal-type' of a research process, and let us understand that different methods are needed to support different phases of a research process and to foster iterative cycles. In the td-net toolbox such a process ideal-type is depicted in the 'search by phases' where we recommend transdisciplinary methods and tools for each phase and its particular challenges. We encourage transdisciplinary researchers to think about when and in which phase they want to craft moments of co-creation, and to specify with whom they envisage interacting and with what degree of intensity (Stauffacher et al, 2008, cited in Lang et al, 2012: 32). In my experience, the (simplified) version of the levels of intensity used by Pohl et al (2017) – to inform, consult or co-produce – is helpful to clarify and discuss expectations with stakeholders. In addition, we often refer to Lang et al's process illustration (see the figure in Extract 3.3) to highlight that the actual doing of the research is only one of the three phases: similar importance should be given to the joint framing of the research as well as to the (re-)integration and implementation of co-created knowledge.

The texts by Chris Rust (Extract 3.4) and Felicity Callard and Des Fitzgerald (Extract 3.1) I read with great interest, as they nicely illustrate the limits of only approaching co-creation with combinations of tools.[6] Tools cannot (or can at least only partially) craft boundary objects, and nor can they protect co-creators confronted by power imbalances or by the requirement to show the unique individual contribution to a co-created research project. The extracts help me to think out of 'my' box and remind me – occupied by my intentions to make co-creation happen – that co-creation is not only and not always beneficial for everyone.

In Extract 3.4, Rust illustrates how Jeff Baggott, filmmaker, and Nick Dulake, industrial designer, artfully reduced (disciplinary-framed) information without losing room for nuances: 'Again, the aim is always to ensure that the visual narrative remains true to the physicists' scientific understanding while being meaningful to the audience. The reason for introducing this account is to show how tacit or visceral communication can be the only way that some knowledge can be transmitted' (Extract 3.4). In my view, this is an example of the high potential of arts and design to craft boundary objects: 'A boundary object permits boundaries between disciplines, scientific fields and practice to be crossed in order to enable common research efforts' (Bergmann et al, Extract 3.2; also mentioned in Lang et al, Extract 3.3). More importantly, Rust reminds me of the importance of tacit transmission: 'Completing the meaning of the work is likely to draw on the tacit knowledge of the viewer' (Extract 3.4). He let me question to what extent we assume that methods and tools lead to co-created knowledge that can be extracted and transferred into the ongoing research process as a separate entity, separated from the co-creators with their embodied knowledge and continuous meaning making. Furthermore, Extract 3.4 explicates how the lack of attribution in co-created products puts the co-creators' standing as researcher and artist at risk – and this seems to affect representatives of the arts and humanities more than representatives of other disciplines. The need for attribution therefore illustrates that the aspired iterative character of co-creation (compare this with Lang et al in Extract 3.3) may put some co-creators more at risk than others, and that process design may also involve the question of when not to co-create.

While the need for attribution may restrain representatives of the arts and humanities from engaging in co-creation in the first place, Callard and Fitzgerald (Extract 3.1) elaborate on power imbalances in ongoing interdisciplinary co-creation. I find these authors' encouragement not to aspire to parity, not to expect direct reciprocity, between natural and social sciences perspectives, refreshing. Not pretending that I have fully grasped their claim, I take the text as an invitation to think more about how we can better understand power dynamics and cleverly act accordingly, and to question whether we want to invest energy in changing power imbalances (Callard and Fitzgerald prefer to invest elsewhere). I have to admit that I still struggle with the idea of giving up the assumption that process design and tools should also contribute to alleviate power imbalances, foster mutual learning and respect, which lead to different – and hopefully more pertinent – co-creation. I take the Callard and Fitzgerald text as a reminder that such an alleviation of power imbalance may only be temporary and does not lever out 'financial and epistemic power' (Extract 3.1).

To sum up, I like the way these excerpts contribute to a multifaceted perspective on co-creating a research project: co-creation can be enabled by a process design that allows for recursive cycles and reflexivity in skilfully composed research teams (Extract 3.3), and by combining integration methods and instruments that enact different dimensions and types of integration (Extract 3.2). The texts also show that co-creation is a time-intensive endeavour, and that you cannot and should not co-create everything. It is helpful to craft moments of co-creative encounters but also to allow moments of deep disciplinary elaboration and to communicate this accordingly. The texts illustrate that co-creation is about acknowledging different perspectives and different ways to contribute to a co-creation process – and also about acknowledging constraints to co-creation and that several constraints affect representatives of different disciplines differently – representatives of the arts, humanities and social sciences existentially (Extracts 3.1 and 3.4).

Notes

[1] The original figure was in colour while the reproduction here is in black and white; the zig-zag line referred to can still be seen in the centre.

[2] I base my comment on experiences made in my position as head of project methods at td-net, where I am responsible for the operation and further development of the online open access td-net toolbox for co-producing knowledge (www.transdisciplinarity.ch/toolbox). The td-net is a competence centre of the Swiss Academies of Sciences (www.transdisciplinarity.ch). It links scientific communities, supports transdisciplinary careers and promotes the development of competencies and methods. Reacting to demands and mandates, I advise research teams regarding method selection, combination and adaptation. As a member of the ITD Alliance Working Group on Toolkit & Methods I am involved in coordinating peer-to-peer exchange with experts in inter- and transdisciplinary method(ology).

3   See also the ITD Alliance Inventory Project at https://itd-alliance.org/inventory-project
4   See the three types of knowledge tool (https://go.transdisciplinarity.ch/TToK) and the
    Outcome Spaces Framework (Mitchell and Fam, 2020).
5   Pohl et al (2017: 43) describe how several transdisciplinary tools can be combined to
    'align their research projects with the requirements of transdisciplinarity' and to 'think
    through ways to better link research to societal problem solving.'
6   This is especially true when talking about situative workshop formats that should open
    spaces in which everybody expresses their perspective on an equal footing and contributes
    to co-creation.

References and further reading

Darbellay, F., Moody, Z. and Lubart, T. (eds) (2017) *Creativity, Design Thinking and Interdisciplinarity*, Creativity in the Twenty First Century Series, Singapore: Springer.

Fleck, L. (1979) *Genesis and Development of a Scientific Fact*, Chicago, IL: The University of Chicago Press.

Lury, C., Fensham, R., Heller-Nicholas, A., Lammes, S., Last, A., Michael, M. and Uprichard, E. (2018) *Routledge Handbook of Interdisciplinary Research Methods*, Abingdon: Routledge.

Mitchell, C. and Fam, D. (2020) *Outcome Spaces Framework*, td-net toolbox profile (9), Swiss Academies of Arts and Sciences: td-net toolbox for co-producing knowledge, https://zenodo.org/record/3717200#. Y8A0SuLP1hE

Pohl, C. and Wülser, G. (2019) 'Methods for Co-Production of Knowledge among Diverse Disciplines and Stakeholders', in K.L. Hall, A.L. Vogel and K. Crowston (eds) *Strategies for Team Science Success: Handbook of Evidence-Based Principles for Cross-Disciplinary Science and Practical Lessons Learned from Health Researchers*, Cham, Switzerland: Springer, pp 115–21.

Pohl, C., Krütli, P. and Stauffacher, M. (2017) 'Ten reflective steps for rendering research societally relevant', *GAIA – Ecological Perspectives for Science and Society*, 26(1): 43–51, https://doi.org/10.14512/gaia.26.1.10

Stauffacher, M., Flüeler, T., Krütli, P. and Scholz, R.W. (2008) 'Analytic and dynamic approach to collaboration: A transdisciplinary case study on sustainable landscape development in a Swiss prealpine region', *Systemic Practice and Action Research*, 21: 409–22, https://doi.org/10.1007/s11213-008-9107-7

Studer, S. and Pohl, C. (in press, 2023) 'Toolkits for Transdisciplinary Research: State of the Art, Challenges, and Potentials for Further Developments', in R.J. Lawrence (ed) *Handbook of Transdisciplinarity: Global Perspectives*, Cheltenham: Edward Elgar Publishing.

4

Funding Collaborative Research

Chapter overview

Science policy is currently experimenting with an increasing push for interdisciplinary and (depending on the national context) transdisciplinary research. Researchers and institutions are called on to conduct more and better inter- and transdisciplinary research and to be open to collaborative research. However, researchers as well as funders confront the challenge of different understandings of inter- and transdisciplinarity in science and policy (see Chapter 1 for a selection of readings on this topic). The decisions that funders make and the intentions behind funding calls have a major impact on how inter- and transdisciplinary research is shaped, the extent of integration, and ultimately its effectiveness. Additional incentives, support for interdisciplinary research leadership, administrative flexibility and a partnership approach between funders and researchers are all required.

The first reading from Flurina Schneider et al (Extract 4.1) introduces us to the roles of 'science policy', which, broadly, encompass research funding, the career pathways of scientists and the organisation of the science–society interface. This article analyses the role of science policy in promoting transdisciplinarity as a way of making research part of societal transformations, and focuses on how research funding programmes could enhance the implementation of transdisciplinary research. Based on a discussion with representatives of four transdisciplinary funding programmes, the authors develop a model that shows the key stages relevant to the enhancement of transdisciplinary research, and offer guidance for the implementation of future such programmes.

Diana Rhoten (Extract 4.2) then provides some historical perspective to the Schneider et al findings, describing how the lack of systemic implementation within universities and other research institutions can impede funders' efforts to promote interdisciplinary ways of conducting research. This early empirical study identifies some of the extrinsic and intrinsic factors in play, and argues for organisational reforms to derive maximum benefit from the funding available to support interdisciplinary research.

The excerpts from Julia Stamm's investigation of the European Commission's Framework Programme for funding research and innovation (Horizon 2020) (Extract 4.3) offer insights into a specific science policy instrument designed to promote the involvement of the humanities and social sciences in interdisciplinary research, which, as we have noted in the 'Introductory Essay', is still predominantly influenced by the science, technology, engineering and mathematics (STEM) disciplines.

EXTRACT 4.1

Schneider, F., Buser, T., Keller, R., Tribaldos, T. and Rist, S. (2019) 'Research funding programmes aiming for societal transformations: Ten key stages', *Science and Public Policy*, 46(3): 463–78.

1.3 The role of science policy for TD research

As the field of TD research has developed, many scholars have pointed out how the prevailing research context shaped by current science policy is persistently unfavourable to TD modes of knowledge production; TD requires conditions that differ from those needed for basic disciplinary research (Dedeurwaerdere 2013; Kläy et al. 2015; Kueffer et al. 2012; Schneidewind 2009). For example, implementation of TD research requires time, skills, and resources for collaborating with other disciplines and societal actors throughout the research process. This process must include efforts towards joint problem framing, exploration of goals and pathways to societal transformations, and co-production and communication of knowledge with and to non-scientific actors. In addition, evaluation of the quality and impact of TD research demands criteria that do justice to the TD character of the project (Roux et al. 2010).

Research funding bodies increasingly acknowledge the importance of TD research, yet their management, evaluation, and funding practices often do not reflect this (Woelert and Millar 2013). For example, there is much evidence that interdisciplinary and TD research proposals have difficulty obtaining funding, since reviewers typically apply disciplinary perspectives and quality criteria instead of considering the integrated whole (Bromham et al. 2016; Mansilla 2006; Woelert and Millar 2013). Moreover, (classic) academic careers are still typically built on measuring scientific impact according to publication in peer-reviewed journals – journals that are more interested in the scientific part of TD research, not in the efforts of such research to contribute to actual societal transformations (Kueffer et al. 2012; Rhoten and Parker 2004). Consequently, for TD research to reach its full potential, experts argue that far-reaching structural and institutional changes are needed in the way academic organizations are managed, organized, and funded (Dedeurwaerdere 2013; Kläy et al. 2015; Kueffer et al. 2012; Schneidewind 2009).

With third-party funding increasingly required for research, research funding programmes and bodies now play a crucial role in science policy (Braun 1998; Bromham et al. 2016; Lyall et al. 2013) and, consequently, in possible changes to the science policy context. Funding bodies strongly influence what kind of research programmes get launched, what research proposals get funded, what kinds of impacts are valued, what networking and capacity-building opportunities are possible, and what sort of career experience is considered valuable in applicants for funding.

1.4 How can TD research funding programmes become more effective?

In recent years, an increasing number of research funding bodies have been implementing entire funding programmes dedicated to addressing diverse societal challenges by means of TD approaches (Hoffmann 2016; Lyall et al. 2013; Wardenaar 2014). However, the people responsible for these programmes often face challenges in designing and implementing structures and processes that enable TD knowledge production at the level of an overall research funding programme (Lyall et al. 2013). Indeed, to date, very little documented experience exists in implementing TD research at the programme level, and very few scientific studies have examined such programmes beyond a focus on individual activities (Bergmann et al. 2005; de Jong et al. 2016; Defila and Di Giulio 1999; Klein 2008; Pohl 2011; Roux et al. 2010).

A valuable exception is the work of Lyall et al. (2013), who investigated the role of funding agencies in creating TD knowledge to promote learning and practical guidance to funders. The researchers identified the following key success factors: identification of the appropriate loci of TD, knowledge integration as deliberate steps throughout the programme, inspiring leadership, active management, learning, and continuity. Moreover, they highlighted the following key aspects for consideration by funding bodies: shaping TD research initiatives, reviewing and evaluating TD research appropriately, building TD capacity, encouraging stakeholder engagement, and ensuring the sustainability of interdisciplinary research.

In addition, several scholars have investigated the management of large TD research programmes (Defila et al. 2006; König et al. 2013)—though without necessarily focusing on funding-related activities specifically. In their look at management, for example, Defila et al. (2006) emphasize that TD collaboration does not occur automatically, but rather must be purposefully initiated, moderated, and accompanied. Further, they observe that the management of TD research programmes must be professionalized, and the people managing these programmes must be supported. In their detailed handbook for programme managers, they identify eight main tasks and recommend their implementation in different project phases: (1) joint goals and questions; (2) integration of research networks; (3) synthesis; (4) joint products; (5) selection of persons and team building; (6) involvement of external actors; (7) internal and external communication; and (8) organization of work.

Our article contributes to these debates by investigating how large research funding programmes can support and implement TD research more effectively. With the term 'TD research funding programme', we mean funding programmes whose basic parameters are predefined. They are generally launched by a funding body with the goal of financing research on a specific topic or conceptual issue.

They embrace several independent research projects with their own sub-goals and methods. However, all involved projects are generally expected to contribute to the overall programme goal(s).

[…]

3.2 Ten key stages for TD interaction

Our collaborative research showed that in the course of a TD research funding programme, 10 key stages are relevant to enable successful TD research. Problem and goal definition (Phase A) occurs in Stages 1, 2, 3, 4, 6 and 8; co-production of knowledge (Phase B) in Stages 5, 7, 8; and contributions to societal transformations in Stages 9 and 10 (and, in part, 8). In the following, we will introduce the stages sequentially; however, as noted above, they may actually overlap or reoccur in a cyclical manner. As we mainly focused on the activities and stages that funding bodies and programme managers can influence, we do not present details or recommendations regarding project implementation (Stages 2 and 5).

To illustrate what can be done not only to implement these stages in a beneficial way, but also to show what kinds of challenges can occur in a specific programme, we present details of the activities and experiences of NRP 61 in Table 2.

Table 2. Example: activities and experiences of the TD Swiss National Research Programme (NRP 61) 'Sustainable Water Management', structured according to the 10 key stages of TD interactions

Key stages	Approaches and experiences of NRP 61
1. Programme preparation	In a formal process, various societal actors—including federal offices, research institutes, and NGOs—proposed water as a topic for a new NRP. The Swiss Federal Council made the final selection of the topic, based on a programme outline elaborated by SNSF. After this decision, the appointed programme president wrote a programme call together with the steering committee. While societal actors had some say in defining the general programme topic, there was relatively little societal involvement in the final definition of specific programme goals and requirements. The selection criteria defined in the programme call emphasized scientific quality, interdisciplinarity and TD, compliance with programme goals, and a focus on application and implementation.
2. Proposal elaboration	Project applicants interpreted the programme call in different ways and integrated societal actors to differing degrees. Some built consortiums with scientific and societal actors, while others organized a joint workshop to identify the most relevant research topics before finalizing the proposal. The majority defined societally relevant questions based on their research experience without explicitly consulting societal actors.

(continued)

Table 2. Example: activities and experiences of the TD Swiss National Research Programme (NRP 61) 'Sustainable Water Management', structured according to the 10 key stages of TD interactions (continued)

Key stages	Approaches and experiences of NRP 61
3. Interactions with applicants	Project evaluation was a two-phase process (pre-proposals and full proposals). The requirement of TD was clearly communicated in the call, and applicants were expected to document societal actors' involvement in the problem framing and the planned research. In some cases, projects were asked to improve their TD designs and societal relevance after their pre-proposals were evaluated.
4. Project selection	Evaluation and selection were conducted by a steering committee supported by external reviewers. The head of knowledge exchange with expertise in TD and a representative of the Federal Administration each had advisory roles. Other than this, non-academic societal actors were not involved. Programme representatives reported that the thematic orientation of the submitted proposals did not fully cover all the intended programme subtopics. Moreover, comparatively few proposals demonstrated sound TD competences and research designs. All in all, less than half of the selected project teams were experienced in TD. However, those applying TD designs could receive funding for senior researchers dedicated to TD work.
5. Research activities	Individual projects varied widely in their intensity of TD collaboration. Projects ranged from rather elementary one-way communication designs to sophisticated TD designs aiming at knowledge co-production on equal footing from start to finish. In this stage, most of research projects became very caught up in their specific research questions. Programme goals were relegated to the background and only identified with to a limited extent. Consequently, many rather disciplinary projects found it difficult to grasp what TD would mean and why they should engage in synthesis activities.
6. Joint agenda setting	After project selection, the steering committee adjusted the programme's thematic lines according to the topics of the selected projects. However, many of the projects were not satisfied with their assigned roles and expected contributions. In response to the demand of project teams, the steering committee initiated an ad hoc bottom-up process to define priority topics, which would be expected to bundle the communication of research results to society. The advisory group comprising societal actors was invited, but very few of them wound up playing an active role. Overall, the importance of this stage was underestimated from a TD perspective. There was no agreed-upon strategy of how to come up with a coherent set of priority topics that would integrate the programme goals as well as the goals of the individual projects. Consequently, many research projects resisted TD collaborations towards achieving the programme goals.
7. Networking and integration	Diverse networking opportunities were organized including annual programme meetings. Knowledge exchange between the involved projects and societal actors was also facilitated through creation of 'social learning videos' on the research of individual projects. Synthesis work was conducted by synthesis teams composed of project researchers and societal actors according to four different thematic lines. Additional funding made the synthesis work possible. The synthesis process was challenging due to people's differing expectations of the process and the leadership.

(continued)

Table 2. Example: activities and experiences of the TD Swiss National Research Programme (NRP 61) 'Sustainable Water Management', structured according to the 10 key stages of TD interactions (continued)

Key stages	Approaches and experiences of NRP 61
8. Interactions with participating projects	Projects were required to report about their TD work as part of their annual reporting. The head of knowledge exchange visited each project and offered advice. A workshop was organized about TD research. Research projects involving seniors experienced in TD knowledge production were very satisfied with these support offerings. However, less experienced research projects felt they did not receive enough support in integrating TD into their specific research. They expressed dissatisfaction, for example, about insufficient feedback on their TD efforts; at the same time, they rarely requested such advice. Very few projects were aware that they could request additional funding for TD activities from the programme.
9. External communication and implementation	The head of knowledge exchange developed a comprehensive communication and implementation concept that was continuously adapted throughout the programme. Each project was expected to engage in communication and implementation activities. The programme supported them in creating videos for communication with societal actors, as well as in preparing practice-oriented articles for professional journals. At the programme level, thematic synthesis booklets were produced and public events were organized, with the latter including workshops with practitioners and a touring exhibition. A limited structured debate was held with all the research-programme actors about envisioned societal transformations, stakeholders involved, and methods that would enable achievement.
10. Programme conclusion and evaluation	The programme concluded with two major communication events, including a marketplace for practice-relevant insights. Afterwards, a handful of activities continued, such as the touring exhibition. Several researchers felt that four years was not enough time to maximize their research results vis-à-vis TD. Several projects reported, for example, that development of trust with local actors required more time and that more interactions would be needed to explore opportunities for implementation of the new insights they elaborated. Others reported that they had enough time to better understand a situation, but that more time would be needed to create an applicable tool. The progress and impacts of the projects were monitored by means of annual reporting. Two assessments were conducted towards the end of the programme to learn about the overall programme experience with TD research: one assessment focussed on the process of synthesis building (Hoffmann et al., 2017); the other focussed on the involvement of societal actors in the participating projects (Schneider and Buser, 2018). An impact assessment is currently slated to occur two years after conclusion of the programme.

1   *Programme preparation (Phase A, programme navigation).* The first key stage involves the preparation of the programme description. In this stage, the overall parameters of future research are set. This program description provides an initial outline of the relevant societal problems, goals, and research questions, as well as available funds. Moreover, by outlining the general research requirements, it also determines the room for manoeuvre applicants will have in designing and implementing TD processes within specific projects. In most cases, the programme preparation includes a formal call for projects. As such, basic aspects of problems and goals are defined by actors who will not be directly involved in the eventual research projects (e.g. the NRPs). In other cases, it takes the form of a proposal elaborated by the eventual researchers (e.g. the NCCRs).

   Considering the importance of the programme preparation phase in defining the scope of the programme, group members stressed that in a TD programme open dialogues must exist from the very beginning to facilitate mutual understanding and joint definition of problems and goals, in particular, with societal actors. At the same time, they mentioned that certain goal formulations can and should be left somewhat open to allow for later concretization through the grantees (see Stage 2). The latter is also important to enable reframing of programme goals based on the final composition of selected research projects (see Stage 6).

   Moreover, when aiming to contribute to societal transformations through the programme, our research highlighted the importance of formulating not only the thematic topics to be addressed in research (scientific goals), but also the societal transformation goals, as well as the methods and pathways/processes to enable such contributions. In addition, programme frameworks should be defined such that they enable favourable conditions for TD research (see all following stages).

2   *Proposal elaboration (Phase A, implementation of projects).* In response to the programme call, research consortiums then write and submit TD research project proposals specifying particular goals, research questions, and methodological approaches. They take into account the requirements and framings of the call, but also typically add new framings that represent their own problem perceptions and interests, as well as those of the societal actors they collaborate with. As a consequence, developed research proposals might not cover all goals mentioned in the programme call.

   The research also revealed that TD consortiums usually require a generous amount of time to frame problems and goals, since they must first reach mutual understandings about societal and scientific knowledge gaps and priorities. This is especially the case when participating actors have no prior relationship and trust must be built. Research funding programmes can tackle these challenges through activities proposed in stages 3, 4, and 6.

3   *Interactions with applicants (Phase A, project support).* Emphasizing the competitive character of the tender process, programme-level actors usually limit their interactions with applicants to clarification of formal requirements, as well as decision letters in case of two-step evaluation procedures. But as the experiences of our learning group members revealed, applicants are often unclear as to what is expected of them

and what they can apply for. Moreover, considering the dominance of disciplinarily education in academia, skills and competences in TD research are still limited among applicants. Hence, group members observed that research funding programmes can considerably enhance the TD quality of the proposals and ultimately the overall programme by supporting the applicants in the proposal writing stage. This support can range from clear communication of call requirements regarding TD and offering preliminary feedback in the case of two-step evaluation procedures up to and including ensuring adequate time and funding (e.g. by providing seed money for TD problem framing), and arranging training in TD methods. Moreover, the design recommendations that were identified for Stage 8 (interactions with participating projects) are also relevant for this very early interaction with applicants.

4   *Project selection (Phase A, programme navigation).* In this stage, the programme evaluates the proposals and decides which projects will be included in the programme, what is expected of them, and how much funding they will receive. This stage is a key in a TD programme, as it determines the final programme composition, including the involvement of all relevant thematic sub-topics and TD competences, as well as the provisioning of adequate funding of TD work.

Thus, in the selection stage, group members highlighted that projects should be chosen that contribute to the programme's defined scientific goals and to its intended societal transformation goals. In addition, consideration should be given to innovative ideas proposed by applicants. But it was also stressed that research programmes should begin by optimizing the framework conditions in which research projects operate to support their TD endeavours. For example, it showed to be crucial to ensure funding of experienced senior researchers and process facilitators who are capable of moderating processes of knowledge co-production and overseeing integration of different types of knowledge. Projects based mainly on PhD research can hardly fulfil the demanding work of TD collaboration due to lack of time, experience, and seniority. It should go without saying that TD work requires its own budgetary allowance and should not be regarded as a free addition. Funding of transformation and follow-up processes often turned out to be required later to bring research results to fruition. However, it was also stressed that if TD designs and efforts appear insufficient, the programme should ask the projects to make improvements to demonstrate that TD is taken seriously.

Our comparison of different programme approaches revealed three basic strategies for selecting projects: (1) selecting a set of individual projects, in which each project applies a systematic TD and transformation-oriented approach while addressing a sub-goal of the programme; (2) selecting a suitable mix of rather disciplinary projects, in which the programme goals are addressed by means of TD synthesis efforts at the programme level; (3) selecting projects that result in a combination of the first two strategies. In the first strategy, decisions about project approval should consider the TD quality of each project proposal and the TD competences of all applicant teams. In the second and third strategy, not every research project must be designed in a fully TD way; however, the composition of selected projects must be appropriate and TD process knowledge must be assured at the programme level.

Whatever the case, in selecting research projects, TD research programmes must apply evaluation criteria and procedures that do justice to the TD character of research proposals.

5   *Research activities (Phase B, implementation of projects).* With the official start of the programme, the approved research projects begin implementing the research described in their proposal. In this stage, TD processes of knowledge co-production are implemented at the level of the individual projects. Researchers collaborate with the societal actors relevant to their specific project questions. Our study showed that implementation of TD research can be very challenging. For example, progressing TD collaboration often requires adaptation of the research designs to respond to needs and perspectives of different disciplines and societal actors. Research funding programmes can substantially support the projects during this stage (see Stage 8).

6   *Joint agenda setting (Phase B, programme navigation).* After the research projects have been selected, a joint agenda-setting stage is needed to redefine problems and goals (Phase A) previously defined independently by different actors (e.g. individual research projects, on the one hand, and the steering committee, on the other). To establish a firm basis for engagement of individual projects in the TD synthesis process, it is important to align or realign the programme's scientific topics and societal transformation goals with the topics and goals set out in the approved research projects. It is also necessary to agree on organization of the TD synthesis process itself, including the involvement of stakeholders. Moreover, it ideally provides an initial opportunity for individual projects to begin taking ownership of programme goals.

This stage is often overlooked in official programme plans. However, our research shows that when this stage is not taken seriously and goals are mainly defined by the steering team of a given TD research programme, frictions can emerge in later phases of the programme and/or programme results may be deemed illegitimate for overlooking the normative or practical concerns of societal actors. One example of how this stage can be designed are the workshops that were organized at the start of the NCCR North–South programme, in which researchers and societal actors jointly identified the main problem areas to be addressed by the different research teams (Wiesmann et al. 2011).

7   *Networking and synthesis (Phase B, programme navigation).* Networking and synthesis activities follow the stage of joint agenda setting. In this stage, insights generated by individual research projects are brought together and project members exchange knowledge. It may be regarded as the core processes of TD co-production of knowledge at the programme level. All the programme representatives involved in our study stated that designing and shaping these TD processes at the programme level is very challenging, since it requires facilitation of TD collaboration between heterogeneous individual research projects and societal actors from different levels. They reported that interaction events not considered meaningful by participants are typically not attended or are subject to criticism. They also reported that shaping TD processes requires particular expertise and authority.

Based on these insights, group members concluded that TD programmes should: (1) elaborate sound TD methodological frameworks in the programme preparation and joint agenda-setting stage, enabling all researchers and societal actors to jointly agree on the themes and processes of TD collaboration; and, (2) that these processes should be implemented and facilitated by experienced TD specialists. The frameworks must outline promising sequences of TD interactions and assign roles to involved actors that the latter consider meaningful, for example, by gaining new insights or clearly seeing that their individual contribution is important and shapes the pathway towards jointly set goals.

8 *Interactions with participating projects (internal communication, training, and monitoring) (Phases B–C, project support).* Supportive interactions with participating projects—including internal communication, training, and monitoring—is an on-going task (see also Defila et al. 2006; Lyall et al. 2013). It can range from providing clarifying information about basic requirements and offering targeted trainings—e.g. for early career scientists—to setting up adequate monitoring systems for evaluation of individual project performances. Ideally, in a TD research programme, these activities are optimized to support the TD work of individual research projects as effectively as possible.

Our study revealed the following two crucial elements: (1) Support researchers in developing and applying knowledge, skills, and competences of TD collaboration. TD competences such as fluency in suitable methods for TD knowledge production or possession of strong social and communicative skills are indispensable to the design and facilitation of fruitful TD research (Herweg et al. 2012; Wiek et al. 2011). However, these skills and competences are seldom taught in typical academic curricula. As a consequence, many researchers involved in our TD research programmes have only limited knowledge and experience regarding TD. They use the term 'transdisciplinarity' in their proposals to respond to the programme calls, but their respective competences and creativity are restricted. To tackle this situation, TD research programmes must find ways of supporting researchers in developing and applying the required knowledge, skills, and competences. One mentioned option is to organize training sessions to familiarize applicants with the methods and requirements of TD research. Subsequently, participating projects can be given access to TD advice, training courses, or peer knowledge-exchange events about integrative methods and communication skills. Annual reporting could be systematically employed on behalf of formative evaluation and provision of feedbacks regarding TD progress—preferably including face-to-face meetings.

(2) Demand and reward TD, transformation-oriented working modes: Since TD competences are not mainstream in academia, any research programme aiming to foster TD research must explicitly demand TD designs and processes from the project applicants. This means that the requirements of TD work must be clearly communicated and that implementation of these designs and methods must be an integral part of the annual reporting. If TD designs and efforts appear insufficient, the programme should ask the projects to make improvements. At the same time, it is equally important to reward TD efforts. Rewarding of TD efforts should begin at the

moment of project selection and continue through the realization of successful TD processes and outcomes. In both cases, it is key that not only scientific publications are counted, but also outcomes that (are likely to) have a transformative impact on society (see Step 9). Also, rewarding TD efforts implies to allow flexibility if ongoing TD collaborations require adaptations of the TD research designs.

9   *External communication and transformations (Phase C, programme navigation/ implementation of projects)*. In a TD research programme, this stage goes beyond classic activities of knowledge transfer such as when research results are communicated to society in a one-way process. It involves more diverse and collaborative forms of interaction such as knowledge exchange, joint learning, and transformative practices (Lang et al. 2012; Mitchell et al. 2015; Roux et al. 2006). Communication and implementation activities usually intensify in the second half of a given programme, and may occur at the level of individual projects and/or the overall programme.

Three modes of promising communication and transformation-oriented activities could be identified: (1) Generation of targeted knowledge products for science and society belongs to the most widespread contributions of our TD research funding programmes. Knowledge products included leaflets, reports, maps, software programs, decision-making tools, or radio/TV broadcasts. The research showed that knowledge products were especially valued by societal actors when they addressed a knowledge gap of their concern and when the products contained the right level of detail. In one case, for example, hydropower companies and government representatives were eager to know how the availability of water resources might change under climate change conditions and how they should adapt their management practices. In this case, it proved highly useful to generate and supply them with graphs visualizing the possible future evolution of available water resources. At the end of the programme, standardized metrics can be used with relative ease to evaluate the generation of targeted knowledge products.

(2) Facilitate learning processes on technical, normative, and practical aspects. Another regularly mentioned key contribution of TD research funding programmes is enhancement of mutual learning processes among researchers and other societal actors. As the diverse societal actors involved in a societal challenge may have differing problem perspectives and priorities, merely identifying and provisioning 'scientific facts' often is not enough to support societal transformations. In many cases, the facts may be contested or may be difficult to translate into practical actions capable of addressing the challenge under scrutiny. In the cases we investigated, for example, this came up in relation to questions about what a 'just' (i.e. fair) water governance system would look like and what the management alternatives might be. In such situations, one important societal goal—as aptly put by Mitchell et al. (2015)—may be to facilitate emergence of 'new perspectives, new orientations, new strategies, and new tools—seeing and doing things differently as a result of their experience of TD research'.

TD research programmes can foster this outcome through the creation of spaces for mutual learning between researchers and various societal actors to reflect on their diverse problem framings, normative assumptions, and the significance of new knowledge for practical actions capable of addressing the societal challenge. Spaces for mutual learning can be created through careful organization and facilitation of workshops, field trips, dialogue events, and informal encounters, as well as through participation in events organized by societal actors (e.g. policy dialogues). The synthesis stage of a given programme is also a particularly suitable stage to emphasize the facilitation of mutual learning.

(3) Strive to improve societal problems. Many of the interviewed programme participants suggested that improving societal problems represents the gold standard of TD research. Depending on the specific problem context, improvements were seen as any of the following: tangible changes in structural obstacles, institutional settings, and management strategies or practices; shifts in policy or societal discourses; spread of more inclusive and participatory forms of collective decision-making; realization of organizational innovations or adaptations; dissemination of technologies or application of decision-making tools. However, other than simply presenting research results, they were often not very explicit about how research can contribute to such improvements. In many cases, research programmes do not last long enough to oversee effective translation of innovative research insights into societal transformations.

As a consequence, group members recommended that TD research programmes more carefully investigate: how changes in the targeted societal field can be brought about; how the research programme can contribute to such changes; and what must be achieved before the programme concludes so that societal transformations might continue to unfold even without the direct involvement of the programme. One solution explored by two participating research programmes was to fund small pilot projects that specifically implement transformative activities as an integral part of the overall programme. Another possible solution mentioned involved more careful planning of follow-up processes and interfaces (e.g. patronage) between the research programme and subsequent societal efforts.

10 *Programme conclusion and impact evaluation (Phase C, programme navigation).* The conclusion of a research programme usually means wrapping up all its activities and communicating its final results. However, contributions to societal transformation often require more time to unfold. Thus, in many cases, follow-up activities are needed to effectively (re-)integrate TD research results into societal practice. Therefore, group members stated that the concluding stage of TD research programmes should also include the handover of responsibility to other suitable actors/organizations capable of carrying on the initiated work as needed.

Moreover, evaluation of scientific and societal impact should be part of all large research programmes to assess their performance and learn for future programme designs (Lang et al. 2012). Impact evaluation should cover the activities of individual participating projects, but also the programme as a whole. Ideally, evaluations will be carried out at different points in time to capture both short- and long-term impacts emerging from the TD research programme. Finally, the relevant TD processes and emerging pathways to societal transformation should also be explicitly addressed in the evaluation.

EXTRACT 4.2

Rhoten, D. (2004) 'Interdisciplinary research: Trend or transition', *Items and Issues*, 5: 6–11.

E.O. Wilson has argued that consilience—the "jumping together of knowledge" across disciplines "to create a common groundwork of explanation"—is the most promising path to scientific advancement, intellectual adventure, and human awareness (Wilson 1998: 8). Wilson and other interdisciplinary advocates contend that the breaching of scientific boundaries will lead to other breakthroughs as critical as the cracking of the DNA code.

Today, some analysts claim that academic science has already embraced the idea of consilience and that a transformation is well underway from the traditional manner of doing research—homogeneous, disciplinary, hierarchical—to a new approach that is heterogeneous, interdisciplinary, horizontal, and fluid (for example, Cooke 1998; Etzkowitz and Leydesdorff 1998; Gibbons et al 1994). Others, however, suggest that the university's metamorphosis toward interdisciplinarity is nowhere as far along as those in the first camp maintain (for example, Hakala and Ylijoki 2001; Hicks and Katz 1996; Slaughter and Leslie 1997). In fact, some would even argue that there is no empirical evidence of any fundamental change encompassing the university science system (Shinn 1999; Weingart 1997).

Our recent NSF-funded study of interdisciplinary research centers and programs suggests that the latter camp is right to be skeptical. Across the spectrum of higher education, many initiatives deemed interdisciplinary are, in fact, merely reconfigurations of old studies—traditional modes of work patched together under a new label—rather than actual reconceptualizations and reorganizations of new research. It was common to hear, for example, the mechanical engineer, atmospheric physicist, and public policy analyst describing themselves as "co-investigators on an interdisciplinary project" yet to observe them conducting their respective pieces of the research in near isolation from one another. Conversely, it was rare to encounter the hydrologist, economist, ecologist, and decision manager "collaborating directly with one another in the field" to formulate a new multi-objective integrative model.

Conventional explanations of the failures of interdisciplinary research to gain traction in the academy typically cite the following factors: the lack of funding for such initiatives; the indifference or hostility of scientists to working across established boundaries; and the incompatibility of university incentive and reward structures with interdisciplinary practices (for example, Bohen and Stiles 1998; Klein 1999; Metzger and Zare 1999; National Academies 1987, 2000; Weingart 1997). While these explanations are not wrong per se, our research suggests that the first two claims may be overstated while the third actually underestimates the broader set and deeper source of organizational misalignments.

By adapting Huy and Mintzberg's (2003) "triangle of change" and applying it to the academic research environment, this article demonstrates that the transition to interdisciplinarity and consilience does not suffer from a lack of extrinsic attention at the "top" or intrinsic motivation at the "bottom," but, rather, from a lack of systemic implementation in the "middle".

[...]

The fact is, universities have tended to approach interdisciplinarity as a trend rather than a real transition and to thus undertake their interdisciplinary efforts in a piecemeal, incoherent, catch-as-catch-can fashion rather than approaching them as comprehensive, root-and-branch reforms. As a result, the ample monies devoted to the cause of interdisciplinarity, and the ample energies of scientists directed toward its goals, have accomplished far less than they could, or should, have.

No systemic implementation

If neither attention nor motivation are lacking in the pursuit of interdisciplinarity, what forces are preventing its promotion from trend to transition? We argue that despite "talking the talk" of cross-boundary collaboration, many universities are failing to "walk the walk." Instead of implementing interdisciplinary approaches from the perspective of a thorough-going reform, many universities are simply adopting the interdisciplinary labels without adapting their disciplinary artifacts. The result has been problematic on two levels. Not only has the persistence of old structures created real or perceived disincentives to and penalties for pursuing interdisciplinary work. But, far more critically, the lack of systemic implementation taken in order to re-design and not just re-name these structures and thus actively support interdisciplinary research has actually created initiatives that are inherently incapable of achieving the very goals they seek to accomplish and unfortunately unable to serve the very constituents they hope to support. Below are just a few cursory examples of some of the common organizational errors that have resulted from the lack of vigorous thinking around interdisciplinarity.

The interdisciplinary centers we studied here, as well as most of those we have since observed, are organized around large catch-all themes such as "global climate change," "environmental impacts," or "sustainable resources." Yet, they often lack unified and unifying problem definitions and project directions around which their researchers' skills and ideas could coalesce. While purposefully broad themes allow a certain amount of disciplinary multiplicity, the absence of explicit, discrete targets of work—otherwise known as "boundary objects"—appears to complicate rather than catalyze communication and collaboration between the disciplines. As a result, most interdisciplinary research centers have a tendency to become a nexus of loosely connected individuals searching for intersections, as opposed to cohesive groups tackling well-defined problems. This result is more akin to the traditional department structure—minus the common ground—than it is an example of a new mode of knowledge production.

Similarly, most centers we examined began by creating a "laundry list" of affiliates and disciplines at the proposal stage, instead of selecting on the basis of the research problem and identifying what researchers might potentially contribute. In combination with a trend in interdisciplinary funding toward longer-term initiatives, this has meant that researchers—having been chosen to fill a nominal slot rather than address a specific role—often find themselves "locked in" to center affiliations from which they do not benefit professionally and may not even thrive intellectually despite their own motivations and interests. In several cases, researchers reported forsaking the extrinsic rewards for the intrinsic ones but in the end getting neither: "I was left with nothing but feelings of frustration and ambivalence with the interdisciplinary center, and feelings of fear and rejection in my disciplinary department." Thus, while longer organizational life cycles may give centers time to improve their research practices and processes, long-term and full-time affiliations can actually limit and not accentuate researcher creativity and productivity. In our study, researchers who felt free to enter and exit collaborative relationships reported more progress with their interdisciplinary projects and greater satisfaction in their professional lives overall.

In the same vein, interdisciplinary centers seem to have associated larger numbers of affiliates with greater rates of interdisciplinarity. While this may make sense in terms of increasing disciplinary multiplicity, our data show that it does not increase meaningful interdisciplinary activity. In fact, our results suggest that although medium and large centers (20-49 affiliates and 50 or more affiliates, respectively) may produce marginally more information-sharing relations within and across disciplines on average than small centers (fewer than 20 affiliates), they are not necessarily more effective at producing interdisciplinary knowledge-creating connections. Indeed, we found that small centers— or small bounded networks within large centers—actually produce more such connections than larger centers do.

Moreover, because many large centers are inter-institutional or international, they must rely on cyber-infrastructure to support interdisciplinary science. While such technologies make long distance science collaborations plausible, the data indicate that technologically-mediated communication may be a good complement but not a good substitute for face-to-face communication—particularly when working across different disciplines. Approximately 71% of the researchers in our study reported face-to-face communication as their primary mechanism for information sharing and knowledge creating, both in general and across disciplines. This compares to 59% who reported using technologically-mediated communication in general, and only about 50% who employ technologically-mediated communication across disciplines. Finally, the fact that 77% prefer informal to formal face-to-face communication in both circumstances reinforces other research suggesting that the sharing of scientific information and the creation of new knowledge are dependent on the interpersonal, spontaneous interactions of researchers (Kanfer 2000)—a class of interaction generally hindered by traditional disciplinary departments and so often unrealized by new interdisciplinary centers.

Some implications and conclusions

At the outset of our study, we were struck by how little empirical data existed about the real-world practice of interdisciplinary research. Two years later, we are struck by the fact that our data raise more questions than they answer. And yet, even so, we believe there are a number of clear implications to be drawn from our study regarding the future conduct of cross-boundary science.

To provide fertile ground for this type of research, interdisciplinary centers need not only to be well-funded but to have an independent physical location and intellectual direction apart from traditional university departments. They should have clear and well-articulated organizing principles—be they problems, products, or projects—around which researchers can be chosen on the basis of their specific technical, methodological, or topical contributions, and to which the researchers are deeply committed. While a center should be established as a long-standing organizational body with continuity in management and leadership, its researchers should be appointed for flexible, intermittent but intensive short-term stays that are dictated by the scientific needs of projects rather than administrative mandates. Not only will such rotating appointments allow researchers to satisfy their intellectual curiosities without jeopardizing their professional responsibilities, they will also better serve the epistemological priorities of interdisciplinary research.

As more researchers divide their time between interdisciplinary centers or programs and traditional disciplinary departments, the academic research community must learn to accommodate institutionally and professionally what Brown and Duguid (2000) describe as "networks of practice." Networks of practice constitute the broad social systems through which researchers share information but which do not always yield new knowledge in immediate or traditional forms. In the current academic structure, the value of research and researcher alike is usually measured by the production of new knowledge in the form of publications in academic journals. However, information sharing networks may often yield "harder to count" but equally important— albeit different—outputs such as Congressional testimonies, public policy initiatives, popular media placements, alternative journal publications, or long-term product developments. While these are the opportunities that often draw individuals to interdisciplinary work, they are also some of the most under-appreciated and unrewarded activities within today's academy.

Finally, for interdisciplinary research centers to achieve their stated aim of addressing new problems in fundamentally new ways, they must be populated with individuals who can serve as "stars" and as well as those who can be "connectors." These are not always one and the same. Universities, therefore, will have to reconsider the priorities and practices of graduate education and training in order to prepare individuals for such centers. We argue that graduate programs must not only educate future scientists to be experts in the methods, techniques, and knowledge of their chosen disciplines but to have the broader problem-solving skills that require learning, unlearning, and relearning across disciplines.

How best to support and encourage these new ways of learning is the central challenge now facing the academy. All around us, the sciences are increasingly colliding at the nexus of complex problems. In the years ahead, those collisions have the capacity to produce many interdisciplinary discoveries as seminal as Watson and Crick's. The universities that successfully reform themselves to meet the challenges presented by interdisciplinary research will find themselves at the center of what some observers liken to a second scientific revolution. Those who fail will find themselves watching from the sidelines.

Stamm, J. (2019) 'Interdisciplinarity Put to Test: Science Policy Rhetoric vs Scientific Practice – The Case of Integrating the Social Sciences and Humanities in Horizon 2020', in D. Simon, S. Kuhlmann, J. Stamm and W. Canzler (eds) *Handbook of Science and Public Policy*, Cheltenham: Edward Elgar Publishing, Chapter 19.

Today, interdisciplinary approaches are broadly being acknowledged as *the* appropriate tool to understand and tackle complex societal issues. There are increasing demands for funding to be organised not in terms of disciplines, but along thematic areas, thus asking for multi-perspective approaches to address them. Budget constraint in research funding also means that funders increasingly focus on research that can generate relevant and implementable solutions (Al-Suqri and AlKindi 2018: 8).

[...]

Interdisciplinary research and collaboration are being hailed as fundamental preconditions for solving the problems and challenges facing societies and our planet. Interdisciplinarity is being encouraged in science policy discourse and among funding agencies. It is considered a central element of funding programmes, such as the European Union's Framework Programme for Research Horizon 2020 (H2020), with its Societal Challenges being designed to overcome disciplinary boundaries and 'silo-thinking' in order to address complex and interdependent problems.

[...]

A large part of research on interdisciplinarity has focused on the institutional obstacles that discourage interdisciplinarity or render interdisciplinary work difficult. At the same time, expectations were formulated that interdisciplinary cooperation can be improved through changes in institutional settings (MacLeod 2018). However, while very often at the centre of attention, institutional barriers are not the only ones that confront interdisciplinary work. In addition to structural issues, power asymmetries, hierarchies and prejudices also play a role in interdisciplinary interactions. Frequently, these asymmetries refer to the roles and status of disciplines in the human and social sciences vis-à-vis the natural and technical sciences. This is shown, among others, by the work of Mathieu Albert and his colleagues in Canada. Focusing on SSH scholars whose experience has 'generally been overlooked by analysts' (Albert et al. 2017: 85), they investigate interdisciplinary collaboration within highly structured academic institutions, with the aim to study the integration of scholars in a research culture for which they were not adequately prepared by their formal education. The authors argue that the challenges faced by SSH scholars in faculties of medicine differ considerably from those faced by SSH scholars engaging in interdisciplinary collaboration while remaining in SSH departments. Looking at the role of social scientists in the Canadian Institutes of Health Research, they find that the interdisciplinary practice in the institute is far from the visions formulated at the creation of the organisation. The study carried out by Albert and his colleagues revealed substantive power imbalances between SSH scholars and clinical scientists, with SSH researchers constantly having to 'explain, demand, fight, and advocate for their own work and that of their peers' (Albert et al. 2017: 98).

Albert's findings regarding the often difficult position of SSH researchers in interdisciplinary research efforts are echoed by numerous other publications (ESF 2013; Delbridge 2015; Felt 2015; Serpa et al. 2017). In her Nature article 'Integration of social into research is crucial', Ana Viseu from the University of Lisbon writes that SSH scholars are often brought into research projects not as scientists in their own right, but to 'maximise the benefits of research' and to mitigate negative impacts and ensure public acceptance (Viseu 2015). She also points out that integration is all too often asymmetrical. SSH scholars tend to be brought in after the project has been defined, and they do not seem to be entitled to sit at the table and contribute equally. On the other hand, natural scientists sometimes accuse social scientists of poor rigour and of dedicating too much time to conceptualising problems (Zic-Fuchs 2018: 323). Forging a shared mission and nurturing constructive dialogue are identified as crucial elements to overcome these tensions. Interdisciplinary collaboration cannot start after the problems are clear, the questions are defined and the methodologies decided upon – collaborative problem-framing is at the very heart of the interdisciplinary research process (Lyall et al. 2011).

[...]

another distinction that helps us to understand interdisciplinarity and its different interpretations consists in identifying whether it is initiated and driven mainly by science managers and policy makers or by scientists themselves. These two approaches are being referred to as managerial top-down interdisciplinarity and practitioner bottom-up interdisciplinarity (Mäki 2016). Many researchers feel that, while bottom-up interdisciplinarity that is generated by the scientific community itself comes naturally, top-down approaches tend to be imposed and artificial and, therefore, are unable to generate the expected outcome. This is reflected in the fact that academic work on interdisciplinarity often perceives it as a top-down phenomenon (Al-Suqri and AlKindi 2018), pushed by science policy makers and research funders. It is doubted that top-down approaches to interdisciplinarity can actually achieve 'true' interdisciplinarity, defined in terms of the integration of ideas, concepts and methodologies from different disciplines. 'When "interdisciplinary" approaches are adopted, largely in response to top-down requirements of funders ... these tend to be multidisciplinary in nature, with team members working primarily within their existing disciplinary paradigms and often involving dual rather than integrated agendas' (Al-Suqri and AlKindi 2018: 10).

[...]

Science policy makers and funders have, therefore, a key role to play in preparing the ground and providing the framework for meaningful interdisciplinarity. But many of them struggle. All too often, interdisciplinarity is considered as a goal in itself, rather than a process that takes time and requires active support, careful steering and thoughtful management. Interdisciplinary collaboration does not just happen; it has to be initiated, planned and continuously revisited. This requires wise leadership, as well as careful expectation management. As mentioned earlier, integration needs to happen from the very early stages of the research process on, not towards its end, to avoid that researchers will be asking different questions in different ways. Including mechanisms for self-reflection and learning, as well as capacity building, contributes to the success of interdisciplinary programmes. While producing interdisciplinary research is not an easy endeavour in itself, evaluating this work is also far from trivial. The evaluation of interdisciplinary work is another highly important and contentious issue that calls for the increased attention of policy makers and funders. Here, the work of Michèle Lamont and her colleagues provides relevant insights. Evaluators in interdisciplinary panels have to 'walk a fine line between trust and creed to produce a fair evaluation' (2006: 53).

Researchers willing to engage in interdisciplinary work desire – and require – more support for their research. The provision of funding alone is insufficient.

[...]

The two reports from 2015 and 2018 (as well as the one that had been published in-between) make very clear that the European Commission's aspiration to 'fully integrate the social sciences and humanities' has so far not been successful – and this despite using a methodology likely to overstate the level of SSH research integration (EASSH 2018). A number of factors can be identified that lead to this finding.

As stated before, interdisciplinarity is not an achievement per se, but needs to be framed and supported in appropriate ways. The current Horizon 2020 approach to fostering interdisciplinary collaboration comes without a strategy. It conveys the impression that interdisciplinarity, once announced as the new goal of research funding activities, simply happens, that calling for more interdisciplinary consortia and integration of the different disciplines will happen automatically. Interdisciplinarity, however, needs to be nurtured. It needs to be accompanied and proactively fostered. Furthermore, it cannot be a one-size fits-all approach, and approaches to realising interdisciplinarity in research will have to vary depending on its context.

The current top-down approach to interdisciplinarity adopted by the European Commission appears to be the result of the conviction of the importance of interdisciplinary efforts to solve Societal Challenges, rather than a reflective process allowing for learning and readjustment.

Commentary

Catherine Lyall

There are at least two distinct ways in which readers of this collection might approach the topic of 'funding collaborative research'. On the one hand, we could consider how research funding bodies design and administer inter- and transdisciplinary research funding schemes. On the other hand, we could look at this from the perspective of a researcher asking, 'How should I go about raising funding for my collaborative research?'

We can find plenty of information to help answer both of these questions in the three extracts included in this chapter. There is sound advice here on the practicalities of inter- and transdisciplinary research funding, and I will reflect a little on this in what follows. There is also much to reflect on in the thought-provoking observations about the power dynamics and politics of such funding.

One thing that immediately struck me was how familiar the challenges that each of the authors addresses were, and, on the surface, how little these challenges seem to have abated in the years between Diana Rhoten's study (Extract 4.2) and the other two. In those intervening 15 years, while there has certainly been much 'vigorous thinking' from the inter- and transdisciplinary scholarly community, institutions have still not responded adequately to Rhoten's criticisms about the lack of systemic implementation. Julia Stamm's observation in Extract 4.3 that 'provision of funding alone is not sufficient' is key and is echoed in Flurina Schneider et al's analysis, in Extract 4.1, of what is – still – required to facilitate effective transdisciplinary programmes.

As a collection, the three extracts particularly highlight for me several aspects of collaborative research funding and, in discussing these, I will also try to pinpoint a few practical considerations for researchers when constructing a budget for an inter- or transdisciplinary grant proposal:

- The duration of such funding
- The flexibility of such funding
- The roles that funders should play
- Functions that funding should support.

Elsewhere (Lyall and Fletcher, 2013), we have written about the challenge of maintaining interdisciplinary scholarship within the context of a publicly-funded research system where money injected as, for example, programme funding for groups of researchers to work on a particular topic area may not garner sufficient institutional support to maintain long-term interdisciplinary scholarship. Creating – and maintaining – interdisciplinary research capacity and infrastructure (including interdisciplinary journals, conferences, teaching programmes, career tracks and physical spaces designed to facilitate interaction) requires sustained investment. Without such support it is hard to achieve major advances in addressing some of the complex, multifactorial problems that we face, such as climate change or social justice. As Schneider et al note in Extract 4.1, structural and institutional changes are needed in the ways academic organisations are managed, organised and funded.

At the same time, in Extract 4.2, Rhoten reminds us of the challenge of retaining interdisciplinary capacity without limiting researcher mobility and the renewal of research agendas.

In Extract 4.1 Schneider et al open with a cogent summary of the challenges of conducting transdisciplinary research and indicate how they think these are influenced by external research funders. They then offer some proposals for how transdisciplinary research funding can become more effective. Importantly, this article makes us think about not just how we run inter- and transdisciplinary projects, but how we manage larger scale programmes. Table 2, included in the extract, is especially valuable in highlighting some of the common challenges and pitfalls when designing and implementing a transdisciplinary research programme.

Duration of funding is a recurrent theme in discussions about inter- and transdisciplinary research; commonly, as in Extract 4.1, researchers report that the typical three- to four-year project grant allows insufficient time to maximise research results. As Schneider et al describe, building trust with knowledge co-producers may require considerable time in new collaborations. Towards the end of a project, more time may also be needed to explore implementation opportunities for research findings. In talking about the need for inter- and transdisciplinary research centres 'not only to be well funded but to have an independent physical location and intellectual direction apart from traditional university departments', Rhoten (Extract 4.2) foreshadows Stamm (Extract 4.3) who says that funding alone is not enough. However, I do struggle with Rhoten's suggestion that we should make appointments to interdisciplinary centres on a rotational basis. While I appreciate her concern that stability and continuity might lead to stagnation, and I understand that she has the best interests of academics' careers in mind, to me, this implies that interdisciplinary research careers are second class, and denotes what we might now characterise as a 'fix the people, not the system' approach.

We could argue, based on some of the things we read in Extract 4.1, that transdisciplinary research requires greater continuity of funding, and that such research is ill suited to time-limited, project-based funding. Here, Rhoten also has something to offer. In noting that academic 'value' is embodied in the production of new knowledge in the form of publications in academic journals, Rhoten is foreshadowing a very current debate about how we measure what she describes as the 'harder to count' but equally important outputs, which she describes as 'some of the most under-appreciated and unrewarded activities within today's academy'. In the UK there are attempts to recognise such 'impact' activities in our national Research Evaluation Framework (REF)[1] that allocates public funding according to institutional performance (see, for example, Smith et al, 2020), but this metric remains problematic for inter- and transdisciplinary research (Lyall, 2022).

In their discussion of the programme preparation phase, I was particularly inspired by Schneider et al's phrase about needing 'room for manoeuvre', and this raises an important issue about the need for greater flexibility in inter- and transdisciplinary research funding. More so than a mono-disciplinary project, interdisciplinary projects may need to develop and change as they respond to the needs and perspectives of different disciplines and societal actors. Over-planning can constrain a project's ability to be responsive to change: while the end goal should be clear, the routes to achieving it might be subject to revision as the project progresses. Yet, we are all too familiar with the highly competitive funding environment and funders' requirements that applicants stipulate precise research designs and justify all associated costs. This increased flexibility would also require a change of mindset in peer review (see also Chapter 5 in this book).

In the case of transdisciplinary research, where relationships among participating actors may only be nascent, Schneider et al (Extract 4.1) caution that proposals may too often be based on academic researchers' experience without input from societal actors. The same is true, these authors note, for the overall programme design, where 'basic aspects of problems and goals are defined by actors who will not be directly involved in the eventual research projects'. Transdisciplinary research can place significant demands on external participating actors, especially those from non-governmental organisations (NGOs), charities or local community groups, and can lead to 'stakeholder fatigue'. Joint agenda setting prior to any funding being available may not be possible. Do we do enough to support the involvement of such societal actors if we want anything more than banal 'letters of support' to submit with our grant applications? This speaks to the need for funders wishing to stimulate transdisciplinary research to allow for seedcorn funding to build relationships that will truly enable joint problem framing, as Christian Pohl describes in his commentary in Chapter 5 (see also Chapter 3 in this book).

In our review of a UK research programme (Lyall et al, 2013), we identified a number of roles for funding bodies, namely: shaping inter- and transdisciplinary research initiatives, reviewing and evaluating inter- and transdisciplinary research appropriately, building inter- and transdisciplinary research capacity, encouraging stakeholder engagement, and ensuring sustainability of this research. Extract 4.1 offers further suggestions on the role that funders can and should play in the development and funding of research programmes while noting that funders are often constrained by the competitive nature of the tender process. Schneider et al also acknowledge that those responsible for designing transdisciplinary research funding are challenged when implementing structures and processes that enable transdisciplinary knowledge production at the level of an overall research funding programme. I found their observation that 'most of [the] research projects became very caught up in their specific research questions.

Programme goals were relegated to the background' certainly resonated with my own experiences as both a participant in, and evaluator of, large-scale inter- and transdisciplinary research programmes.

The selections from Stamm's chapter (Extract 4.3) reinforce many of the messages from Schneider et al in terms of the practicalities and responsibilities of funders (including the European Union, which is the focus of her chapter), and the key role funders play in preparing the ground and providing the framework for meaningful interdisciplinarity. Stamm is especially useful (see the sixth paragraph in Extract 4.3) in reminding us that *who* initiates interdisciplinarity (in all its forms) is significant whether it is initiated and driven mainly by science managers and policymakers or by researchers. This can have a bearing on what such research can achieve: whether it is indeed truly interdisciplinary or essentially multidisciplinary in nature. This problem is echoed by Schneider et al when they attribute the lack of transdisciplinary skills and competences to 'typical academic curricula', noting that many researchers involved in the transdisciplinary research programmes they assessed had only limited knowledge and experience regarding transdisciplinarity. Thus, if we are to achieve value for money from the funding invested in inter- and transdisciplinary research, funders of such research programmes must invest more in developing and supporting the required skills and competences, including potentially more coaching for applicants drafting their proposals, if we are to avoid the consequences that Stamm identifies.

Rhoten's paper (Extract 4.2) draws our attention to the 'stars and connectors' within inter- and transdisciplinary research. I would argue that we are good at funding – and rewarding – the former, but not the latter. Despite evidence to the contrary (Hoffmann et al, 2022), this essential integration function (also highlighted by Schneider et al) risks being viewed as a 'soft skill' rather than a core research role that underpins success in inter- and transdisciplinary research. The integration role is something that applicants should foreground and cherish when drawing up their budget for research support. Too often, this vital function is relegated to a 'communications' role and not awarded the status and funding that it merits within a collaboration – as Schneider et al testify, 'shaping TD processes requires particular expertise and authority'. Failure to recognise this results in, as these authors say, interaction events not being considered meaningful by participants, resulting in poor attendance and critical feedback. Such interaction often needs to carry on beyond the lifetime of a project grant, so, as Schneider et al recommend, the concluding stage of inter- and transdisciplinary research programmes should also include the handover of responsibility to other suitable actors and organisations capable of carrying on the implementation phase, which further speaks to the need for continuity of funding.

Finally, what about readers who want to know more about the tactics of budgeting for inter- and transdisciplinary research? Schneider et al (Extract 4.1) have many practical suggestions for things that funders can do and issues that researchers should address in their proposals. Stamm (Extract 4.3) prompts me to think about the drivers and, consequently, power dynamics of inter- and transdisciplinary research funding, and Rhoten (Extract 4.2) raises issues about the longer-term implications of such funding. Drawing on my own experiences as an interdisciplinary academic (and these span roles in research development, as a research coordinator and manager, and as an independent researcher and evaluator), I glean several tips from the readings offered in this chapter, of which the following seem key:

- Take advantage of any seedcorn funding to build relationships with societal actors in order to facilitate joint problem framing.
- Secure an adequate budget for start-up and implementation costs; be realistic about what you can achieve within a prescribed funding limit, and never assume that networking costs are discretionary and dispensable.
- Foreground the research integrator role, and ensure it is funded appropriately.

Note
[1] www.ref.ac.uk

References and further reading

Graf, J. (2019) 'Bringing concepts together: Interdisciplinarity, transdisciplinarity, and SSH integration', *fteval Journal for Research and Technology Policy Evaluation*, 48: 33–6, https://repository.fteval.at/id/eprint/433

Hoffmann, S., Deutsch, L., Klein, J.T. and O'Rourke, M. (2022) 'Integrate the integrators! A call for establishing academic careers for integration experts', *Humanities and Social Science Communications*, 9: 147, https://doi.org/10.1057/s41599-022-01138-z

König, T. and Gorman, M.E. (2017) 'The Challenge of Funding Interdisciplinary Research: A Look inside Public Research Funding Agencies', in R. Frodeman, J.T. Klein and R.C.S. Pacheco (eds) *The Oxford Handbook of Interdisciplinarity* (2nd edn), Oxford: Oxford University Press, Chapter 36.

Lyall, C. (2022) 'Excellence with Impact: Why UK Research Policy Discourages "Transdisciplinarity"', in B. Vienni-Baptista and J.T. Klein (eds) *Institutionalizing Interdisciplinarity and Transdisciplinarity: Collaboration across Cultures and Communities*, Abingdon: Routledge, Chapter 2.

Lyall, C. and Fletcher, I. (2013) 'Experiments in interdisciplinary capacity-building: The successes and challenges of large-scale interdisciplinary investments', *Science and Public Policy*, 40(1): 1–7, https://doi.org/10.1093/scipol/scs113

Lyall, C., Bruce, A., Marsden, W. and Meagher, L. (2013) 'The role of funding agencies in creating interdisciplinary knowledge', *Science and Public Policy*, 40(1): 62–71, https://doi.org/10.1093/scipol/scs121

Smith, K., Bandola-Gill, J., Meer, N., Stewart, E. and Watermeyer, R. (2020) *The Impact Agenda: Controversies, Consequences and Challenges*, Bristol: Policy Press.

5

Evaluating Interdisciplinary and Transdisciplinary Research

Chapter overview

Disciplines help to organise knowledge for quality assessment purposes and are the cornerstone of academic peer review and recognition. How do we judge what 'good' research is if it is outside of our immediate sphere of expertise? The selected readings in this chapter cover the topic of evaluation, from the general principles (Extract 5.1) to specific questions that help to prosecute the process of assessing inter- and transdisciplinary proposals (Extract 5.4).

Evaluation remains one of the least-understood aspects, Julie Thompson Klein (Extract 5.1) argues. This first reading constitutes a foundational work on the topic of evaluation of collaborative practices. The article synthesises seven generic principles that provide a useful framework to think about evaluation, building recommendations based on a literature review and the author's experience. Alongside detailing these principles and their suitability for such a complex task, the author reflects on changing connotations of the underlying concepts of discipline, peer and measurement.

Different understandings of inter- and transdisciplinarity make it difficult to evaluate projects and programmes that use these approaches. At the same time, evaluation processes shape and transform science and scientific agendas. In Extract 5.2, Katri Huutoniemi and Ismael Rafols deal with this problem by offering an account of what evaluation of interdisciplinarity involves. They provide tools for measuring three properties of interdisciplinarity (breadth, integration and transformation) based on quantitative methods of analysis.

Building on the UK's long experience in evaluating interdisciplinarity, Tom McLeish and Veronica Strang draw conclusions on an in-depth examination of a range of interdisciplinary projects and the work of a UK-based working group of funders and researchers (Extract 5.3). Five areas of evaluation (publishing, research grants, careers, interdisciplinary research centres, institutions) demonstrate both commonality and difference in the task of measuring the added value in interdisciplinary research collaborations. This is a relevant contribution that is suitable for other contexts and countries, while highlighting the main obstacles shared in different contexts when evaluating collaborative research initiatives.

To overcome obstacles in evaluation processes, the final reading (Extract 5.4), written by Christian Pohl et al (2011), discusses the state of the art of the topic based on the experience of the Network of Transdisciplinary Research (td-net, Switzerland) and the academic literature. The authors suggest a number of questions to evaluate inter- and transdisciplinary research proposals, emphasising the quality of synthesis and integration. They provide useful guidelines on how to conduct such an evaluation by complementing the standards assessing the disciplinary quality of proposals.

EXTRACT 5.1

Klein, J.T. (2008) 'Evaluation of interdisciplinary and transdisciplinary research: A literature review', *American Journal of Preventive Medicine*, 35(Supplement 2): S116–S123.

Conclusion: The Logic of Discipline, Peer, and Measurement

An emergent literature is a benchmark of both what is known and what remains to be known. Key insights from this literature appear in Table 2. Yet findings are still dispersed across multiple forums, even with systematic efforts to disseminate information by groups such as the Europe-based td-net.[25,26] Longitudinal empirical studies of interdisciplinary and transdisciplinary evaluation remain few in number and need testing in local contexts. Access to in vivo deliberations is still limited in peer review, and governments lack clearly defined and tested criteria for prioritizing funding across the spectrum of disciplinary and multidisciplinary–interdisciplinary–transdisciplinary research. And, more broadly, unquestioned assumptions about three underlying concepts—*discipline, peer,* and *measurement*— continue to cloud the discourse on evaluation.

Disciplines provide crucial knowledge, methodologies, and tools for interdisciplinary and transdisciplinary work. However, in many discussions, disciplines are still treated uncritically as monolithic constructs. Studies of disciplinarity reveal that disciplines exhibit a striking heterogeneity, and that boundary crossing has become a marked feature of contemporary research. Some disciplines, Vickers[13] observes, have undergone so much change that characterizing them as stable matrices with consensual evidentiary protocols is problematic. Some new interdisciplinary and transdisciplinary fields also reject

Table 2. Key insights

Principle number	Evaluation principles	Key insights
1	Variability of goals	**Variances:** size, scope, scale, level and subsystem, degree of integration in multidisciplinary–interdisciplinary–transdisciplinary environment
		Multiple goals: for example, epistemologic or methodologic forms, product development, pragmatic problem solving *Range of stages:* ex ante, intermediate, ex post
2	Variability of criteria and indicators	**Two major approaches to quality assessment:** conventional metrics; indirect, field-based, and proxy criteria vs primary or epistemic measures of warranted interdisciplinary knowledge in the substance of the work
		Expanded indicators: for example, experimental rigor, aesthetic quality, new explanatory power, feedback to multiple fields, enhanced research capabilities, changing career trajectories, new public policies and treatment protocols, long-term impacts and unforeseen consequences
3	Leveraging of integration	**Key factors:** balance in weaving perspectives together into new whole, reaching effective synthesis, antecedent conditions for readiness **Criteria for leveraging and evaluating integration:** organizational, methodologic, and epistemologic components; strategies that promote communication and consensus; generative boundary objects
4	Interactions of social and cognitive factors in collaboration	**Requirements:** for example, calibrating separate standards, managing tensions among conflicting approaches, clarifying and negotiating differences among all stakeholders, compromising, communicating in ongoing and systematic fashion, engaging in mutual learning and joint activities
5	Management, leadership, and coaching	**Requirements:** managing tensions in balancing acts, consensus building, integration, interaction, common boundary objects, shared decision making, coaching the process **Categories of leadership tasks:** cognitive, structural, and processual
6	Iteration and feedback in a comprehensive and transparent system	**Requirements:** attuning a pluralism of values and interests, iterative work to insure collaborative inputs, transparency to include common stakeholding, feedback to the mission in a dynamic framework, mobility of participants, interaction and communication patterns
7	Effectiveness and impact	**Expanded indicators:** sensitivity to variety of goals in Principle 1 and variety of criteria and indicators in Principle 2; inclusion of unpredictable long-term impacts, returns on investment, value-added

disciplinarity in whole or in part, and, Sperber[10] observed in an online virtual seminar, the purpose of interdisciplinary work may aim to undermine current understanding in disciplines. A standard assessment procedure can help in charting a program's interactions within a broader environment and ensuring that work is sound and reliable.[24] Yet stringent evaluation criteria for both research and evaluation may be counterproductive, especially, Langfeldt[5] warns, for risk taking and "radical interdisciplinarity." Conflicting assumptions about quality meet head-on during *peer* review, whether in ex ante evaluations of grant proposals and priority setting in national research systems or in ex post assessments of research performance and outcomes. A "commonly agreed yardstick" must be developed to "moderate the conservative forces" of traditional research communities, safeguarding against bias.[5]

Identifying experts who fit the "problem space" is crucial, because they form an appropriate interdisciplinary epistemic community. The task is more difficult, though, in emerging fields where the criteria of excellence are not defined yet and the pool of qualified experts is often smaller. In highly innovative work, developing validation criteria to gauge progress often becomes part of the actual process of inquiry.[7] The summary report[2] of the 2006 AAAS symposium cites a number of strategies in funding agencies, including creating "on-the-fly" electronic review teams, using "interpreters" who bridge the epistemic gap among content experts, asking candidates for grants to contribute the names of suitable peers, and forming joint panels and "matrix" schemes that combine disciplinary reviews with full-panel reviews among discipline-based and interdisciplinary members. Special funding programs may bypass conventional control mechanisms, but they run the risk of marginalizing interdisciplinary and transdisciplinary research.[2]

Lamont and colleagues' study[8] of fellowship competitions in social sciences and humanities furnishes a powerful analytical lens for thinking about interdisciplinary and transdisciplinary evaluation. Building on the work of Max Weber and Emile Durkheim, the team described the production of legitimacy that occurs in review panels. Review panels are "sites where new rules of fairness are redefined, reinvented and slowly recognized."[8] In the absence of customary rules, consensus on what constitutes a good proposal must be negotiated. Equilibria must be achieved between the familiarity and distance of non-expertise, between transparency and opacity, expertise and subjectivity, and between interdisciplinary appeal and disciplinary mastery. Methodologic pluralism is key to arriving at a judgment that is both consistent and limits bias.[8]

Finally, the logic of measurement returns the question of evaluation full circle to the gap between conventional metrics and the complexity of interdisciplinary and transdisciplinary research. Paralleling interdisciplinary studies and learning assessment, interdisciplinary and transdisciplinary research process and evaluation are grounded in the philosophy of constructivism. Appropriate evaluation is made, not given. It evolves through a dialogue of conventional and expanded indicators of quality. Traditional methodology and statistics have a role to play, but they are not sufficient. In the past, Sperber[10] admonishes, people seeking the legitimization of interdisciplinary initiatives had to be both parties and judges, educating their evaluators in the process of doing and presenting their work. The emergent literature provides both parties and judges with an authoritative portfolio of methodologies, instruments, design models, guidelines, and conceptual frameworks anchored by a growing body of case studies and findings. They neither impose nor forestall evaluation awaiting a single-best or universal method that would be antithetical to the multidimensionality and context-specific nature of interdisciplinary and transdisciplinary work. They facilitate informed definition of the task and credible tracking of the actions and outcomes attendant to the substance, constitution, and value of the research.

EXTRACT 5.2

Huutoniemi, K. and Rafols, I. (2017) 'Interdisciplinarity in Research Evaluation', in R. Frodeman, J.T. Klein and R. Pacheco (eds) *The Oxford Handbook of Interdisciplinarity* (2nd edn), Oxford: Oxford University Press, Chapter 35.

RESEARCH evaluation (or quality assessment) means the systematic determination of the merit, worth, and significance of a research activity. It implies the existence of both a judgment of quality and a set of organizational procedures and outcomes that are associated with that judgment (Brennan 2007). The evaluation of interdisciplinary research, however, is a tricky issue. On the one hand, the concept of interdisciplinarity already contains a presupposition that it is a valuable thing, as it offers something that is missed by disciplines. On the other hand, because of its deviation from disciplinarity—one of the conditions of possibility of academic knowledge—its value is somehow dubious. This chapter addresses this dilemma by discussing the expected benefits of interdisciplinarity in knowledge production, and the means and measures by which those benefits may be acknowledged and captured in research evaluation.

Such normative consideration of interdisciplinarity requires a critical awareness of the two principal meanings of the term "discipline": first, it refers to a particular branch of learning or body of knowledge, and second, it refers to the maintenance of order and control (Moran 2002). This echoes the relationship between knowledge and power, which is crucial for considering questions of research evaluation. Evaluation is, in essence, a means of exercising control over knowledge. In the case of interdisciplinary research, however, there is no consensus on the legitimate sources and types of control over it. Underlying the debate are uncertainties that center on the concept of interdisciplinarity itself, which often comes in different variations, such as multi-, inter-, cross-, and transdisciplinarity.

First, it is far from clear what defines the quality (or excellence) of interdisciplinary research. Whenever research crosses boundaries between disciplines, the problem arises that each discipline carries specific and sometimes conflicting assumptions about what constitutes quality. The criteria of disciplinary communities are proving insufficient for research that expands, integrates, or challenges the discipline's own canon. In such intellectual exchanges, what exactly is it that determines the relevant criteria: one's own discipline or the other discipline, or some combination of the two—or perhaps knowledge users outside of academia? Second, and related to the previous point, it is unclear who judges interdisciplinary work. Since there is no clearly defined community of peer reviewers as there generally is for disciplinary research, competent reviewers can be very hard to find. Thus, peer review is often biased toward established approaches, unreliable in assessing interdisciplinary work, and helpless in comparing different types of excellence against each other (see Holbrook, this volume). Third, there is no agreement on what constitutes interdisciplinarity, and how it can be identified and measured in practice (Huutoniemi et al. 2010). The definitional debate tends to be paralyzed by the notion that interdisciplinary research can have so many profiles (see Klein 2006, 2008).

The present chapter seeks to address these issues by performing a meta-analysis of the concept of interdisciplinarity in research evaluation. We aim to overcome two major divisions in the existing understanding of the topic. First, there are different normative framings of interdisciplinarity, which shape assumptions about quality and how it is best determined. From this discussion, we identify three major epistemic values or guiding principles of interdisciplinarity and discuss their meaning for research evaluation. Second, there is a growing gap between conceptualizing and measuring interdisciplinarity in research evaluation and the purposes these endeavors have come to serve. Rather than prioritizing one approach over the other, our aim is to bring them together in a mutually reinforcing way. Parallel to the conceptual discussion of the values of interdisciplinarity, we provide bibliometric approaches for mapping and measuring the cognitive properties of research that can be associated with those values. The combination of qualitative and quantitative definitions makes the chapter particularly useful for the purposes of reconsidering and designing research evaluations from an interdisciplinary point of view. The actual implementation of these definitions is likely to differ between particular evaluative settings and is thus beyond the scope of this chapter.

35.1. Epistemic Values of Interdisciplinarity

Evaluations are used, among other things, to certify research activity as valid, to distribute resources in academia, to improve the performance of researchers and organizations, to inform strategic decisions, and to legitimate scientific knowledge in society. Different functions of evaluation raise different questions about interdisciplinarity and offer different kinds of control over knowledge production. In order to better understand and deal with these issues, we review some of the main benefits or "goods" that interdisciplinarity is expected to convey. We are not so much offering a procedure of evaluating interdisciplinary research as giving an epistemic account of what would be involved in doing so.

Interdisciplinarity is not an end itself, but a means of advancing knowledge. To this end, it has several assets that are not, or not appropriately, provided by disciplinary research. In scholarly and policy discussions on the epistemic benefits of interdisciplinary research, three overarching values stand out: breadth, integration, and transformation. Following the standard usage of terms (e.g., Klein, this volume), one might classify research pursuing these values as multidisciplinary, interdisciplinary, and transdisciplinary, respectively. However, our aim is not to provide specific evaluation criteria for different categories, but to illuminate the various "added values" that may and do span those categories. While the primary focus of this chapter is on interdisciplinarity as an academic endeavor, the epistemic values are also relevant for research that involves actors beyond the academic realm. In what follows, we discuss the meaning, implications, and relevance of these values for research evaluation, and summarize them in Table 35.1.

Table 35.1. Three Major Epistemic Values of Interdisciplinarity and Their Implications for Research Evaluation

	Breadth	Integration	Transformation
Value added	Expanded repertoire of specialized expertise	Synthesis of perspectives	Transformation of specialized worldviews
Accountability	Multiple disciplines	Integrative research context	Hybrid communities, future generations
Evaluative focus	Management of diversity	Integrative process	Creativity, renewal of knowledge structures
Epistemic standards	Combination of disciplinary standards	Specific standards for integration	Proactive, emergent standards
Policy implications	Structural flexibility in the evaluation process	An evaluation system of its own	New governance of knowledge production
Proponents	Academic organizations, sociologists of science	Problem-oriented organizations, practitioners, and theorists	University reformers, antidisciplinary movements
Pathologies	Increase of bureaucracy, lack of community	Institutional isomorphism with disciplines, including their limitations	Epistemic anarchy, no cumulative advancement

EXTRACT 5.3

McLeish, T. and Strang, V. (2016) 'Evaluating interdisciplinary research: The elephant in the peer-reviewers' room', *Palgrave Communications* 2, article 16055.

Towards solutions—creating an evaluative framework for Levels (2) and (5)

In this section, we illustrate how the joint summative and detailed evaluatory approach to IDR may be developed in practice, by focussing on two of the five levels identified in Section I: the interdisciplinary grant proposal (Level 2) and the national research evaluation exercise (Level 5).

The few suggested frameworks for evaluation of IDR proposals have an interesting commonality in form: they have generally attempted to identify the holistic structures of good IDR through the formulation of questions to address in relation to the proposal, project or output. For example, both Lyall and King (2013), and Strang and McLeish (2015) condense their findings into a "checklist". We have indicated **in bold** in the box below the questions proposed by Lyall and King that appear to be specific to the evaluation of the fundamentally integrative nature of IDR.

1   Does the proposal describe clear goals, adequate preparation, appropriate method, significant results, effective presentation, reflective critique?
2   How was the problem formulated?
3   **How diverse are the disciplines, methods and researchers and how suitable is the combination of disciplines?**
4   **Is there a clear justification for the choice of disciplines based on the needs of the research questions?**
5   Is the study sufficiently anchored in relevant literature?
6   What is the relationship with the methodology?
7   How will communication be tackled?
8   **Does it describe how the insights of the disciplines involved will be integrated (in the design and conduct of the research as well as in subsequent publications) and how this relates to the type of interdisciplinarity involved; does it demonstrate how the quality of integration will be assured?**
9   **How is the collaboration organized—is there an understanding of the challenges of interdisciplinary integration, including methodological integration, and the "human" side of fostering interactions and communication, and an effective strategy to achieve this?**
10 Is the leadership role and management strategy to deliver the desired outcomes clearly articulated?
11 **Do the researchers involved have demonstrable interdisciplinary skills and experience?**
12 **In particular, is there evidence of interdisciplinary leadership?**
13 Is there an appropriate plan for stakeholder/user engagement from the outset of the project?
14 Does the proposal budget for, and justify, the additional resources needed?
15 **Is it clear how interdisciplinarity will be reflected in the project outputs and outcomes?**

The questions **in bold** seem to lean towards the assessment of the integrative and emergent, and in the case of strongly IDR may require particular expertise. However, such questions could equally be addressed to any research proposal or output where "disciplines" might be replaced by "integrated knowledge" or "methodologies". Furthermore, if they are included in such an evaluation, they enhance the quality of scrutiny given to even a single-disciplinary research programme. This transferability is supported by the working group report from the Durham IAS (Strang and McLeish op. cit.). The report avers that:

With the recognition that IDR represents a foundation, rather than a superstructure, in the organization of knowledge (for a historical perspective see Weingart in Frodeman et al., 2012), it is evident that:

- principles that guide good IDR can also serve as guidelines for good disciplinary research;
- approaches to evaluation that work well for IDR may usefully inform evaluations of single-disciplinary research (2015: 6).

The observation that such transferability does not work reciprocally is the central reason for the challenge addressed in this article. When the starting point for evaluation is that of single-discipline research, attempts to add special "bolt-on" criteria for IDR can be awkward. But if a holistic, interdisciplinary perspective is assumed from the beginning, then there is no point at which special criteria need to be inserted into an evaluatory scheme. Disciplinary and interdisciplinary evaluatory frameworks do not commute.

The Durham IAS report is also couched in the form of a checklist—a rather large one as separate frameworks of probing questions are derived for each of the levels of evaluation. But these detailed lists are generated from an overarching set of criteria, reproduced in the box below:

1   Is the emergent whole of the IDR greater than or different from the sum of its parts? Do the ingredient disciplines do more than work in parallel but interact, communicate, recombine? Are they sufficient?
2   Is the leadership structure characterized by inclusivity, facilitation, transparency of roles and an equality of contributing disciplines in terms of voice and status?
3   Are additional resources and time planned for dialogue, co-learning and integration between the contributing disciplines?
4   Is it clear how the individual disciplines may benefit on their own terms by engaging with the IDR, noting that this can be transformational?
5   Is there a disciplinary hospitality between the researchers, and to external participants, which avoids a hierarchical view of the contributing disciplines?
6   Are there ways of supporting the social cohesion of the collaborators (recognizing that interdisciplinary support structures may help)?
7   Have the different scales, and communication between them, been recognized in the structure of the research?
8   Are there processes for cohering the different data in the research, quantitative and qualitative, recognizing the need for translation where this is necessary?
9   Is the necessary experience with IDR represented by the team and the leadership as well as training and development in place?
10 Are research plans sufficiently open and flexible to adapt to new questions or directions that might arise unforeseen at the outset?
11 If there are "service disciplines" identified in the research, has this been driven by the project needs and not by assumed prevalence of one discipline over another?

Note that there is a strong correspondence between the key points in the two frameworks we have summarized above, correspondences identified by listed point in Table 1.

The framework proposed by Belcher et al. (op. cit.), also drawn from a wide survey of the literature, is rather different in form. These authors focus on "transdisciplinarity", identifying such research as having "explicit goals to contribute to 'real world' solutions and strong emphasis on context and social engagement" (Belcher et al., 2015: 1). There are some—possibly tangential— questions about a supposed division between research and a putative "real world", and about the notion that "trans" disciplinarity possesses a more practical and applied focus than "inter" disciplinarity. But their list of criteria is useful, in that it is more general and more universally applicable to the evaluation of research quality in general. Thus it aims to assess (1) Relevance, (2) Credibility, (3) Legitimacy (this heading contains much of the special requirements of healthy IDR explicit in the "checklists") and (4) Effectiveness (also comprising aspects such as training and development with IDR in mind). These, too, can be mapped onto the cross-corresponding classes of evaluative criteria from Lyall and King, and Strang and McLeish (Table 1).

King (2008) extracted seven perspectives, or "principles" in evaluation of IDR from her comprehensive review. These were (in her specific definitions): (1) variability of goals; (2) variability of criteria and indicators; (3) leveraging of integration; (4) interaction of social and cognitive factors in collaboration; (5) management, leadership, and coaching; (6) iteration in a comprehensive and transparent system; and (7) effectiveness and impact.

Meshing these four approaches (a rather comprehensive set, as they include reviewed work themselves), reveals a strong emergent classification of evaluative criteria. Together these draw on structural, epistemological and participative aspects of entire IDR projects to articulate powerful sets of guiding questions. We have labelled these criteria sets (see Table 1) as Holistic, Social, Experience, Leadership And Effectiveness. The way these break down into particular guidelines at the five levels identified in the introduction, is specific to each of those levels. We indicate how that process might develop below in the two cases of research grants (Level 2) and institutional review (Level 5).

EXTRACT 5.4

Pohl, C., Perrig-Chiello, P., Butz, B., Hirsch Hadorn, G., Joye, D., Lawrence, R., Nentwich, M., Paulsen, T., Rossini, M., Truffer, B., Wastl-Walter, D., Wiesmann, U. and Zinsstag, J. (2011) 'Questions to evaluate inter- and transdisciplinary research proposals', Berne, Switzerland: Network for Transdisciplinary Research (td-net).

4 Questions for evaluation

In the following a number of questions are proposed to evaluate inter- and transdisciplinary research proposals. The questions are targeted to a reviewer who is asked to assess a proposal for its specific inter- and transdisciplinary qualities, emphasizing integration and synergy. The suggested questions draw on Table 3 and are summarized in Table 4.

The first challenge mentioned in Table 3 is the broadness of the approach. Broadness means the number and the diversity of disciplines, methods and analytical scales (molecular, cellular, individual, social, global) that are combined in an inter- or transdisciplinary project. As depicted in Figure 1, this broadness is one of the means to further distinguish proposals within the purpose of fundamental understanding (Stokols et al., 2003; Mitrany and Stokols, 2005) and of problem solving (Hirsch Hadorn et al., 2006; Hirsch Hadorn et al., 2011). Hence, a first question is

Q1   How divers[e] are the disciplines, methods, scales of analysis and possibly the social actors involved?

The question supports classifying proposals according to their specific purpose. Whereas a high diversity is a sign of inter- and transdisciplinarity, a low diversity – a project combining a small number of disciplines, methods, analytical scales and societal actors alike – may be a sign that a project is not inter- or transdisciplinary research at all. A second question related to broadness is thus

Q2   If the project has a low diversity, does it still fall into the category of inter- and transdisciplinary research?

The second challenge mentioned in Table 3 is integration, the widely accepted core challenge of inter- and transdisciplinary research (Klein, 2008b; Pohl et al., 2008; Jahn et al., 2006; Bammer, 2005; van Kerkhoff, 2005; Huutoniemi et al., 2010). The measures and questions gathered in Table 3 basically centre around the questions of whether there is a good reason for an integrative approach, whether the applicants have concrete ideas (methods, tool) about how to make the integration happen and how the proposed integration is assessed in terms of suitability and a balanced weaving of perspectives. Three corresponding questions for evaluation are:

Q3   How innovative and how suitable is the combination of disciplines and fields of expertise for the specific purpose?

Q4   How elaborate is the approach to integration?

Q5   How balanced is the weaving of disciplines or fields of expertise?

The third challenge is reflection and learning. Reflection and learning is a challenge since the combination of disciplinary perspectives and maybe perspectives of social actors requires – compared to a collaboration of researchers with a similar background – much more time for mutual understanding and learning, and possibly a reframing or adjustment of the planned integration and collaboration. The questions in Table 3 basically propose to ask for the readiness and preparedness for learning and adaptation:

Q6   How elaborate is the approach to self-reflection and adaptation?

The challenge of learning and reflection is specifically relevant for the purpose of reflection-in-action. The purpose of reflection-in-action is to critically review a current practice in order to change it. Therefore the success of the project strongly depends on relationship of those who reflect and those who act. Assessing the quality of reflection-in-action requires thus an answer to the following question:

Q7   How likely is the project to substantially interrelate reflection and action?

The challenge of problem solving in Table 3 is discussed with regard of whether the problem requires an inter- or transdisciplinary approach, whether the problem and the projects contribution are sufficiently elaborated and whether the project and its participants are considered to be able to make a change. As the first of those questions is already discussed in Q3, the further questions for evaluation are

Q8   How elaborate is the problem and the project's specific contribution to its solution?

Q9   How likely is the project to make a substantial contribution to problem solving?

Finally the questions on management, social and leadership skills ask whether the management structures seem to match and support the project's goals and whether the involved researchers and partners will be open minded, if necessary ready to fundamentally revise their scientific perspective, able to collaborate in possibly changing hierarchies and to bear and manage tensions. Two last questions for the evaluation are thus

Q10   How well do the management structures match and support the project's goal and combination of disciplines and fields of expertise?

Q11   How do you assess the applicant's collaborative skills (open-mindedness, self-reflection, dealing with changing hierarchies, ability to bear and manage tensions)?

5 Further use of questions

The questions to evaluate inter- and transdisciplinary research proposals are based on the assumption that an evaluation has to combine existing disciplinary standards with new standards. The new standards – as suggested in Table 4 – have to assess the specific quality of synthesis and integration (cf. Tab. 1, page 2).

Table 4. Suggested questions for evaluating inter- and transdisciplinary research

No	Question	Effect
Assessing broadness to further classify proposals		
Q1	How diverse are the disciplines, methods, scales of analysis and/or social actors involved?	The question further distinguishes the purpose of a project within the classes "fundamental understanding" and "problem solving". A high diversity is typical for "comprehensive under- standing" and "wicked problem solving".
Q2	If the project has a low diversity, does it still fall into the category of inter- and transdisciplinary research?	The question asks the reviewer to reconsider the projects classification as inter- and transdisciplinary research based on the projects diversity.
Assessing integration		
Q3	How innovative and how suitable is the combination of disciplines and fields of expertise for the specific purpose?	The question asks the reviewer to assess originality and suitability of the combination of disciplines and fields of expertise for the specific purpose.
Q4	How elaborate is the approach to integration?	The question assumes that an elaborated understanding of the problem and of the projects contribution to its solution is a sign of high inter- and transdisciplinary quality.
Q5	How balanced is the weaving of disciplines or fields of expertise?	The question assumes that an integration that balances disciplines or fields of expertise is a sign of high inter- and transdisciplinary quality.
Assessing reflection and learning		
Q6	How elaborate is the approach to self-reflection and adaptation?	The question assumes that planned stages of learning and self-reflection and the possibility to adapt the project based on this is a sign of high inter- and transdisciplinary quality.
Q7	How likely is the project to relate reflection and action?	The question asks the reviewer to assess whether the project will connect reflection and action.
Assessing problem solving		
Q8	How elaborate is the problem and the project's specific contribution to the problems solution?	The question assumes that an elaborated approach to integration is a sign of high inter- and transdisciplinary quality.
Q9	How likely is the project to make a substantial contribution to problem solving?	The question asks the reviewer to assess whether the project will support problem solving.
Assessing management, social and leadership skills		
Q10	How well do the management structures match and support the project's goal and combination of disciplines and fields of expertise?	The question assumes that an elaborated management structure is a sign of high inter- and transdisciplinary quality.
Q11	How do you assess the applicant's collaborative skills (open mindedness, self-reflection, dealing with changing hierarchies, ability to bear and manage tensions)?	The question assumes that applicants who are committed to core values of an inter- and transdisciplinary ethics are a sign of high inter- and transdisciplinary quality.

Inter- and transdisciplinary research are means to certain ends or purposes. Hence, each evaluation of inter- and transdisciplinary research should start with the purpose. In section 2 three such purposes were distinguished: Fundamental understanding, reflection-in-action and problem solving. For the former and the latter a further distinction can be made depending on the diversity of disciplines, methods, analytical scales and societal actors involved (Figure 1, Q1 Q2).

Table 5 summarizes the question we suggest for evaluating the quality of inter- and transdisciplinarity depending on the purpose. The questions on integration (03, 04, 05), on the elaboration of the approach towards learning (Q6) and on management and social and leadership skills (010, 011) can be used for all purposes. For the purpose of reflection-in-action (07) as well as for problem solving (Q8, Q9) the possible impact can be assessed in addition.

Table 5. Questions for evaluation depending on the purpose
(Questions in bold address the respective purpose).

Purpose	Fundamental understanding		Reflection-in-action	Problem-solving	
	Detailed	Comprehensive		Wicked	Instrumental
Suggested questions for evaluation	Q3, Q4, Q5, Q6, Q10, Q11		Q3, Q4, Q5, Q6, **Q7**, Q10, Q11	Q3, Q4, Q5, Q6, **Q8, Q9**, Q10, Q11	

The questions are suggestions from the perspective of practitioners and theorists of inter- and transdisciplinary research. They are, however, not ready for being implemented as such in a form for reviewers. Depending, amongst others, on the specific targets of the reviewing panel, on the panel's understanding of inter- and transdisciplinary research, on its current practice of reviewing and on pragmatic considerations, a panel should select the most suitable questions and if necessary reformulate them. In other words, the proposed questions are – considering the state of inter- and transdisciplinary research and its evaluation – of an exploratory character. Their use and implementation should be seen as an experiment that has to be observed in order to learn about the questions and to fine-tune them.

Commentary

Christian Pohl

In my opinion, a scholar who has to evaluate inter- and transdisciplinary projects and thinks 'How do I judge what "good" research is if it is outside of my immediate sphere of expertise?' is not ready for this task. Not to say that I don't know this feeling. However, the readings of this chapter, as well as the existence of the scholars who have written them, indicate that in the meantime we have reached another state of discussion. If you ask an experienced and reflective inter- or transdisciplinary scholar to evaluate interdisciplinary or transdisciplinary projects, (s)he will know what aspects of a project (s)he has to scrutinise in order to judge its quality in interdisciplinarity or transdisciplinarity. Such a scholar will not consider it her/his responsibility to assess the projects' quality in discipline x, because this is not the expertise (s)he brings to the evaluation. Katri Huutoniemi and Ismael Rafols (Extract 5.2) comment that 'evaluation is, in essence, a means of exercising control over knowledge'. This can also be read as: it is in the inter- and transdisciplinary evaluators' power and responsibility to distinguish good from poor inter- and transdisciplinary proposals and projects in order to further develop this kind of joint knowledge production.

I agree with Huutoniemi and Rafols that breadth, integration and transformation are three key aspects of inter- and transdisciplinarity and its quality. The first question to clarify is, however, always (and always here really means always) the purpose. Inter- and transdisciplinary research are a means to a purpose (Boix Mansilla, 2006: 19). The adequate kind of breath, integration and transformation depends on this purpose. As a consequence, the plurality of understandings of inter- and transdisciplinary research is not a problem, but reflects the fact that inter- and transdisciplinary approaches are geared towards different goals (Extract 5.1). The heterogeneity of understandings is, in fact, an asset (Vienni-Baptista et al, 2022), and the first thing an evaluator has to assess is how well a project's inter- or transdisciplinary approach fits the project's purpose. To be able to do so, an evaluator has to be aware of the different understandings of inter- and transdisciplinary research. (S)he has to know at least some of them so that (s) he can argue why a particular approach fits very well to the project's purpose or what aspects are missing that another approach would cover. Furthermore, only once the purpose is clear can the evaluator know what 'suitable' refers to in questions such as 'How diverse are the disciplines, methods and researchers and how suitable is the combination of disciplines?' (Extract 5.3). Or in terms of breadth: 'What breadth and what combination of disciplines and representatives of further parts of society does the project provide? What breadth would be suitable to reach its purpose (for example, to solve a particular societal problem, to develop a specific device, to elaborate a reductionist understanding of an issue into an encompassing one)?'

Integration and transformation are fundamental to inter- and transdisciplinary research. This is also why Bammer (2005) and Bammer et al (2020) characterise inter- and transdisciplinarity by the terms 'integration and implementation sciences'. Both particularities are good entry points to look for differences in quality of projects. Formulated indirectly, the question is, 'Do the researchers involved have demonstrable interdisciplinary skills and experience?' (Extract 5.3). More directly, focusing on integration and transformation or implementation, the questions to ask are:

- Do the project leaders have a proper plan for how to achieve integration and/or transformation?
- Do they allocate enough intellectual, social and financial resources to implement the plan?
- Do they reserve adequate time and resources to repeatedly reflect on how the plan works and to adapt it?

For integration, for instance, good inter- and transdisciplinary proposals will explicitly state what the applicants mean by integration (for example, reaching consensus, relating differing viewpoints), and why this is the adequate kind of integration given the project's purpose. Good projects will furthermore propose a (stepwise) procedure of how to achieve integration. They will be aware of and able to name particular challenges of integration and use methods or tools to address these challenges. They will mandate a powerful person or group in the project to take care of integration. They will see integration as a process that requires attention and to be taken care of from early on. Poor proposals will see integration as something that happens by itself, as a natural by-product of scholars working on the same topic. In terms of integration, they will propose to run (not further specified) 'synthesis workshops' towards the end of the project. They either plan to hire an external moderator for that workshop or mandate a PhD student or a postdoc with the integration task. Poor proposals furthermore do not show awareness of the magnitude of the effort and engagement integration requires or of the theories and methods that are there to conceptualise it and to navigate its challenges.

Good and poor transformation follows the same lines. Good inter- and transdisciplinary projects present a proper plan of how to have impact, for instance with a Theory of Change (Belcher et al, 2020). They think about who they will include when in the project, in what form, and for what purpose (Krütli et al, 2010). They build in ample time to reflect on why the plan and the Theory of Change do not work as expected and to adapt it over the running time of the project. They also think about who to involve in this reflection and why. Poor projects organise an outreach workshop towards the end of the project, targeting either decision makers or the public at large, not further specifying who exactly they mean by these categories, why these are the relevant addressees, and why they will be interested in the results.

A further challenge of inter- and transdisciplinary evaluation is 'getting the process right' (Boix Mansilla et al, 2006: 70). From the point of view of an evaluator, evaluation is ideally a process during which evaluators can give feedback to the project teams who then can react, explain their thinking or further develop the proposal. From an applicant's viewpoint, a longer evaluation process also comes with investment of time and resources without any guarantee of success. This is why I think a laborious stepwise inter- and transdisciplinary evaluation process must be funded. In particular, I think of the first step of transdisciplinary research, the joint problem framing, within which a joint topic has to be selected and framed, an adequate team has to be built and the plan for the project has to be specified. We were able to implement such a first step in the funding scheme #ConnectingMinds of the Austrian Science Fund (Fonds zur Förderung der wissenschaftlichen Forschung, FWF). Projects that passed the first evaluation step were funded with €10,000 to meet and jointly frame the problem and further detail the project. Thus, projects that did not pass the second step of evaluation would at least have these costs covered. The problem framing workshops were also interesting from the perspective of the evaluator. The way project teams organised the event, the way they implicitly or explicitly ascribed roles and responsibilities, as well as how the meeting advanced the project and the full proposal, all this was helpful information to check whether impressions from reading the proposal were correct or needed to be revised.

The second element I consider key in getting the process right is a hearing, where members of the project team and evaluators meet in person (or virtually). Q&A allows evaluators to dig deeper into key elements of a project, and for the applicants to make clear how familiar they are with the challenges of inter- and transdisciplinarity and how to approach them. As an evaluator, I like hearings not only for this efficient way of addressing open questions, but also to get an impression of the project team. Are the roles in the proposal similar to the ones in the hearing, or is there a 'secret boss'? Do the applicants ask whether practice partners can attend the hearing? If yes, how do practice partners describe their role in the project? If no, are there good reasons why practice partners must not be present? Apart from the evaluation, it might also be interesting for practice partners to see what questions evaluators are interested in, and what pressure researchers have to cope with before they get a final funding decision.

Finally, getting the process right also means balancing inter- and transdisciplinary expertise with disciplinary or field-specific expertise. There is no right balance. The balance depends on the goals or purpose of the institution that conducts the evaluation. Is it a basic science-driven funder that wants to make funding schemes a bit more inter- or transdisciplinary? Or is the goal of the funder to run an ideal inter- and transdisciplinary evaluation process and to select projects that are at the cutting edge of inter- and transdisciplinary research? Two ways in which this balance of expertise can be fine-tuned are (1) by the number of evaluators from each side and (2) by the power or weight given to the experts of both types in designing and running the evaluation process. In the FWF funding scheme

#ConnectingMinds, for instance, inter- and transdisciplinary scholars were in the driver's seat. We advised the FWF in planning the evaluation process, we suggested inter- and transdisciplinary reviewers, and we took decisions as a jury. Experts of the respective field also assessed the disciplinary excellence of the proposals. This discipline-based assessment was one input to the evaluation process and also influenced our final decisions. However, it was not a case where disciplinary excellence comes first, and on top of it we then had to select excellent inter- and transdisciplinary projects. Disciplinary excellence was one among several criteria, and the jury made an overall assessment of how well projects fulfilled all criteria. Another funder with another selection goal might give more weight and power to the disciplinary evaluators and less to inter- and transdisciplinary evaluators or try to balance both. As Klein states: 'Appropriate evaluation is made, not given. It evolves through a dialogue of conventional and expanded indicators of quality' (Extract 5.1). The main thing to keep in mind is that it must be adequate to the purpose of the evaluation.

References and further reading

Bammer, G. (2005) 'Integration and implementation sciences: Building a new specialization', *Ecology and Society*, 10(2): article 6.

Bammer, G., O'Rourke, M., O'Connell, D., Neuhauser, L., Midgley, G., Klein, J.T., Grigg, N.J., Gadlin, H., Elsum, I.R., Bursztyn, M., Fulton, E.A., Pohl, C., Smithson, M., Vilsmaier, U., Bergmann, M., Jaeger, J., Merkx, F., Vienni-Baptista, B., Burgman, M.A., Walker, D.H., Young, J., Bradbury, H., Crawford, L., Haryanto, B., Pachanee, C.-A., Polk, M. and Richardson, G.P. (2020) 'Expertise in research integration and implementation for tackling complex problems: When is it needed, where can it be found and how can it be strengthened?', *Palgrave Communications*, 6(5), DOI:10.1057/s41599-019-0380-0.

Belcher, B.M., Davel, R. and Claus, R. (2020) 'A refined method for theory-based evaluation of the societal impacts of research', *MethodsX*, 7: 100788, DOI:10.1016/j.mex.2020.100788.

Boix Mansilla, V. (2006) 'Assessing expert interdisciplinary work at the frontier: An empirical exploration', *Research Evaluation*, 15(1): 17–29, https://doi.org/10.3152/147154406781776075

Boix Mansilla, V., Feller, I. and Gardiner, H. (2006) 'Quality assessment in interdisciplinary research and education', *Research Evaluation*, 15(1): 69–74, https://doi.org/10.3152/147154406781776057

Krütli, P., Stuaffacher, M., Flueler, T. and Scholz, R.W. (2010) 'Functional-dynamic public participation in technological decision-making: Site selection processes of nuclear waste repositories', *Journal of Risk Research*, 13(7): 861–75, DOI:10.1080/13669871003703252.

Vienni-Baptista, B., Fletcher, I., Lyall, C. and Pohl, C. (2022) 'Embracing heterogeneity: Why plural understandings strengthen interdisciplinarity and transdisciplinarity', *Science and Public Policy*, 49(6): 865–77, https://doi.org/10.1093/scipol/scac034

Wang, Q. and Schneider, J.W. (2020) 'Consistency and validity of interdisciplinarity measures', *Quantitative Science Studies*, 1(1): 239–63, https://doi.org/10.1162/qss_a_00011

6

Communicating Interdisciplinary and Transdisciplinary Research Findings

Chapter overview

Learning how to communicate about research with other societal partners is a key skill, although transdisciplinary (and on occasions interdisciplinary) research adopts a co-creation approach that goes much further than simply conveying research findings to potential end users. What, then, are the factors that hinder or support communicating inter- and trans-disciplinary research findings? What do successful inter- or transdisciplinary publishing or transfer processes look like? Because a more profound understanding of this process is still needed, communication and co-creation of findings are discussed in this chapter to provide an introduction of the main topics at stake.

In Extract 6.1, Christoph Kueffer et al detail and analyse the constraints that transdisciplinary researchers and practitioners face in order to publish their research results. Although this article was written in 2007, it addresses the main questions for scholars and journal editors when dealing with the publishing process of inter- and transdisciplinary research. These authors discuss possible steps for creating a more sophisticated and strategic approach to the role of publishing in the development of transdisciplinary research as an important and legitimate research endeavour.

With a focus on the science–policy interface, Christian Pohl argues in Extract 6.2 that transdisciplinary research constitutes a useful means of bridging science and policy. Pohl studies transdisciplinary projects and identifies two types of projects. In the first type, researchers reorganise knowledge according to the interests of the audience. Transdisciplinary research of this type requires a clearly defined audience. Researchers in projects of the second type initiate a co-production of knowledge during which the different policy cultures interact. In this case, transdisciplinary research contributes to policies that have to be developed using a collective process involving multiple policy cultures. Extract 6.2 illustrates one type of communication in which researchers and stakeholders need to take into consideration the policy setting in which their outputs will be (potentially) used. Although Pohl analyses transdisciplinary projects, we argue that his findings are useful to a wider audience dealing with impact in collaborative projects.

In Extract 6.3, Erin Leahey et al discuss how interdisciplinarity impacts on researchers' productivity. As discussed in Chapter 4 in this book, national and international funding agencies and universities worldwide promote inter- and transdisciplinary research as a means for transformative science. Applying a quantitative approach to study scientific publications, the authors in Extract 6.3 assess whether such research is indeed beneficial for a research career, and which costs accompany the potential benefits. Results show a pervasive tension by which researchers are penalised due to spanning disciplines as the resulting research is perceived as less rigorous or of lower quality. This fact affects how knowledge produced in inter- or transdisciplinary settings can be disseminated or communicated. It is also empirically shown in this article that lower productivity is associated with cognitive and collaborative challenges of interdisciplinary research or hurdles in the review process.

EXTRACT 6.1

Kueffer, C., Hirsch Hadorn, G., Bammer, G., van Kerkhoff, L. and Pohl, C. (2007) 'Towards a publication culture in transdisciplinary research', *GAIA – Ecological Perspectives for Science and Society*, 16(1): 22–26.

What is Needed for an Effective Publication Culture in Transdisciplinary Research?

We present three propositions for enhancing a publication culture in transdisciplinary research. They synthesise our assessment of the current options for publishing transdisciplinary research and outline strategies for facilitating the publication of transdisciplinary research in the future.

Proposition 1: Transdisciplinary research needs a combined strategy of disciplinary extension ("endogenisation") and transformation ("specialisation")

In our review and classification of journals that publish transdisciplinary research, we identified two classes of journals: those with roots in the established disciplines and those which aim to transcend disciplines, which we call "discipline-based" and "transdisciplinary-focus", respectively. Our first proposition is that both classes of journals are important for the development of transdisciplinarity. Journals that are discipline-based but concentrate on publishing applied and multidisciplinary research, with some inter- and transdisciplinary papers, are more likely to be read by disciplinary experts. This allows insights from transdisciplinary research to feed back on and trigger innovation in disciplinary science. We use the label "endogenisation" of transdisciplinary research into disciplinary arenas for the process whereby transdisciplinary thinking becomes integrated into, and ultimately endogenous to, existing disciplines.

Transdisciplinary-focus journals, in contrast, open up thinking beyond the constraints of disciplinary boundaries and can foster new forms of creativity. We use the label "specialisation" of transdisciplinarity for establishing transdisciplinary research as a research field in its own right that requires specialised research skills, methods, and ongoing methodological development.

Transdisciplinary research is therefore both a valuable extension of disciplinary research – i.e., traditional disciplines profit from insights gained along new research pathways – and a special kind of research that can transcend established disciplines – i.e., new research fields emerge from transdisciplinary research which are not closely linked to traditional disciplines (cf. figure 3). This dual view of transdisciplinarity is one that we have not seen argued elsewhere.

Proposition 2: Fostering the publication of transdisciplinary research in discipline-based journals requires active engagement of both journal editors and transdisciplinary authors

To realise the potential benefits of publishing transdisciplinary research in discipline-based journals, such journals need to be more attractive to transdisciplinary researchers as an outlet for their work, and the published transdisciplinary research must be relevant to the disciplinary readership. Achieving this requires a combined strategy involving both journal editors and authors.

Journal editors could target transdisciplinary research through special sections or special issues that focus on how the relevant discipline(s) and transdisciplinary research can learn from and strengthen each other. Editors of discipline-based journals may profit because transdisciplinary research is often an underexploited resource of innovative scholarship that potentially has a wide and long-term scientific impact.

To complement such approaches by journal editors, transdisciplinary scholars need to plan from early on in a transdisciplinary research project how their research interlinks with classical disciplines, and keep track of the disciplinary relevance of their transdisciplinary work. By being particularly attentive to how their transdisciplinary research advances disciplinary thinking they can produce relevant, interesting, and readable papers that will appeal to a disciplinary audience and enhance the disciplinary journal's standing. In this regard it is necessary to clarify if a particular innovation of transdisciplinary research contributes to problem-oriented or research practice knowledge (cf. figure 2).

Proposition 3: There are gaps in transdisciplinary-focus journals, which, if filled, will facilitate robust transdisciplinary development and the growth of a "college" of transdisciplinary peers

There are specific gaps in the available range of transdisciplinary-focus journals, which is likely to impede the development of innovative, high-quality, and widely applicable transdisciplinary practice. Most of the transdisciplinary-focus journals are problem-oriented, and we particularly note the dearth of practice-oriented transdisciplinary-focus journals. While this is evident from a quick glance at the number of journals listed in each category in the table, the disparity becomes even more stark when we point out that we have presented only a selection of problem-oriented journals, but we have listed all the practice-oriented transdisciplinary-focus journals we are aware of. The lack of practice-oriented journals strongly suggests that there are few options for pushing the boundaries of transdisciplinary research practice, either through sharing lessons learned or proposing and debating new methodological developments. This also makes it harder for transdisciplinary researchers to find each other and to establish a "college" of peers who build on and critique each other's work, in the process raising the quality and applicability of transdisciplinary methodologies.

There is considerable scope for establishing transdisciplinary-focus journals that set methodology agendas and lead the field. For example, there is as yet no high-ranking *Journal of Transdisciplinary Research Practice* which would provide scope for publishing broadly on transdisciplinary research methods, and promote interaction across a wide range of transdisciplinary researchers. We note, however, that previous attempts to establish broad-scope practice-oriented transdisciplinary-focus journals, such as *Issues in Integrative Studies*, do not seem to have been particularly successful, whereas narrow-scope journals such as *Systemic Practice and Action Research* seem to be flourishing. Closer examination of their histories, along with evaluation of how well these journals meet their aims, is likely to provide valuable lessons for establishing new publication outlets.

EXTRACT 6.2

Pohl, C. (2008) 'From science to policy through transdisciplinary research', *Environmental Science & Policy*, 11(1): 46–53.

2. Analytical framework: interacting policy cultures

In transdisciplinary research and in boundary organizations researchers and stakeholders from diverse sectors of society meet and exchange information. Such exchange must take into account, that each of the sectors – science, the private sector, public agencies and civil society – organizes knowledge and action according to individual time scales, categories, priorities, etc. Each sector of society is a separate social world (Star and Griesemer, 1989, p. 388) or culture, characterized by specific norms, knowledges, practices and discourses

(Jasanoff and Wynne, 1998, pp. 16–18; Miller, 2001, p. 485). Members of each culture may look at the same situation and come to opposite conclusions of what is, and what has to be done. Based on their cultural norms, knowledges, practices and discourses they focus on the various elements of the situation and interrelate and interpret them differently. Limoges (1993, p. 420) speaks of "worlds of relevance", instead of social worlds or cultures, to place emphasis on alternative cultural reference systems as a potential source of controversy.

In transdisciplinary research members of the different cultures interact to co-produce knowledge. Elzinga and Jamison (1995, pp. 575–577) and Elzinga (1996, pp. 226–229) distinguish four such interacting cultures for analyzing science policy agendas, and global climate change research: the bureaucratic, the academic, the economic and the civic policy culture. The policy cultures are "competing for resources and influence, and seeking to steer science and technology in particular directions" (Elzinga and Jamison, 1995, p. 575) and are each characterized by their policy; the bureaucratic culture is concerned with effective administration, coordination, and organization; the academic culture seeks to preserve autonomy, integrity, objectivity and control over funding and organization of science; the economic culture is interested in transforming scientific results into successful innovations to be diffused in the commercial marketplaces; and the civic policy culture is concerned with the consequences and implications of developments in science and technology.

Van Kerkhoff and Lebel (2006, p. 470) propose that the bridge between science and policy be conceptualized as an arena: "This allows us to point to specific instances where research based knowledge and action are interacting but without necessarily implying that those interactions are simple or straightforward". Jasanoff and Wynne (1998, p. 17) depict knowledge production as a system of the four interacting policy cultures (Fig. 1). The term "system" means that the emphasis of an analysis shifts from the policy cultures to the way they interact (Jantsch, 1972, p. 103). The academic policy culture – keen to preserve its autonomy and integrity –mainly informs the other cultures. It serves the economic culture by technology transfer, it "speaks truth to power" (Price, 1965; Wildavsky, 1987) to the bureaucratic culture and it informs the civic culture about new scientific insights in order to enhance public understanding of science. From the perspective of the academic policy culture, the other cultures try to restrict its autonomy. The economic policy culture influences the academic by making research more market-driven and by insuring projects and programs are evaluated according to the economic profit they promise. The bureaucratic culture steers the academic culture by policy-driven research programs but also by setting limits to science, as in the case of genetic engineering. The civic culture influences the academic culture by "enlarging civic participation or articulating underexposed points of view" (Wachelder, 2003, p.264).

The interactions depicted in Fig. 1 are "ideal-typical" simplifications in Weber's sense (1973). They stress particular characteristics of the policy cultures, and the way they relate to each other. This over-simplified form has no direct equivalent in reality. The simplification is motivated by the need to analyze knowledge production in transdisciplinary research. If transdisciplinary research is a process of co-production of knowledge, then research will go beyond the role of providing information and the academic policy culture must find ways to interact with the other cultures and their policies.

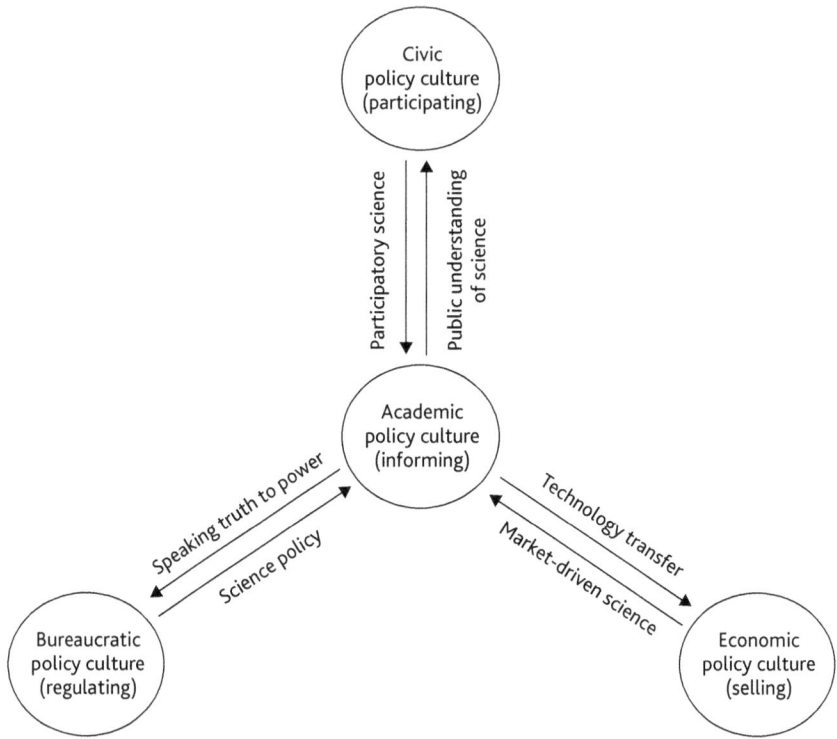

Fig. 1 – Knowledge production as a system of interacting policy cultures (based on Elzinga and Jamison, 1995; Jasanoff and Wynne, 1998, p. 17).

[...]

4. Results and interpretation

For the purposes of analysis the interview passages that dealt with interacting policy cultures were considered. According to the analytical framework this is the case when the researchers go beyond merely informing the other cultures and begin to take on further tasks. Two types of modules can be distinguished. Researchers in projects of the first type remain close to the academic culture's policy to inform, but add reorganizing knowledge as a further task. Researchers in projects of the second type become participant facilitators of a co-production of knowledge involving multiple policy cultures.

4.1. Type one: reorganizing knowledge

Several researchers agreed, in principal, that passing on information was the core task of the academic policy culture. In addition they were concerned with reorganizing knowledge from different disciplinary fields and presenting it to the audience. The audience and its interest is to some extent imagined, meaning that representatives of the audience were not involved in the planning of the modules and the results of the project do not respond to the explicitly formulated demands of the audience.

A first module of type one – "community" – analyzed strategies to induce sustainable development at the community level. Researchers trained in disciplines like economics, psychology, sociology or law studied the potential of regulatory and economic instruments, communicative and diffusive instruments, collaborative agreements and changes in infrastructure as means to induce social change. It was the program's management that merged these projects into one module. Before, the researchers assumed that each project had to report its results individually. The fact that they became one module initiated an internal discussion process:

> Since we all deal with strategies and instruments, we have to develop a common understanding of what we mean: What is an instrument? What is a strategy? What are the different instruments and strategies? Which of those am I studying in my project? (Social scientist)

The discussions yielded a "typology of tools", which was, according to the module leader, the main result of the module's synthesis work. The typology presented the results of the disciplinary projects as alternative instruments, like a toolbox for community mayors. The closing workshop of the module took place in one of the communities that had been studied. The researchers performed a role-play on stage. On the left a community representative sat at his desk; on the right sat all the researchers. Their job was to convince the community representative that their approach to the problem was unquestionably the best solution: that the way forward was through economic incentives, decrees, structural changes or education by advertisement. After a period of trading the conclusion was that a combination of all these instruments would probably best fit the particular context of all the communities.

The academic and the bureaucratic policy cultures were involved in the "community" module. The academic policy culture informed the bureaucratic policy culture. Additionally, the academic policy culture reorganized the findings in a synthesis process, presenting them as tools that could be used by community representatives. From the user's perspective such a synthesis process may seem self-evident and not worth mentioning. Those who have been involved in a project with researchers from different disciplines (e.g. economics, psychology and sociology), will know the long and intensive discussions that are needed before everyone will accept their own disciplinary perspective as one amongst others and to relate the disciplinary results to a practical purpose (Giri, 2002; Loibl, 2005).

A second module of type one – "soil" – developed biological soil remediation technologies. At one site the capacity of plants to extract heavy metals from contaminated soils was studied. At another oil-polluted site, crude-oil-decomposing micro-organisms were applied. The natural scientific projects of the module explored new technologies. The social scientific projects considered the assessment of alternative technologies.

As the assessment group came from a multi-criteria decision making background, the members wanted to conduct a survey in which they would ask the residents for their preferences about alternative soil remediation technologies. Before the survey began a dossier (based on the research group's area of expertise) was distributed to the residents informing them, amongst other things, about the risks of contaminated soil and about uncertainties. The researchers thought of this as neutral information, but the dossier alarmed not only the residents, who suspected the community authorities of downplaying the real danger of the situation, but also the governmental agency responsible for contaminated soil and its remediation and for informing the public. As a consequence of this situation a further survey, intended to compare the residents' attitude towards the oil-eating micro-organisms and their attitude to genetically modified organisms, was prevented by the agency. A natural scientist who was involved understood the reason for the ban: such a survey would "wake sleeping dogs", by making people suspect the researchers of having already tested genetically modified organisms without telling them.

The role of the academic policy culture in the "soil" module was primarily to inform. In addition, an assessment of alternative soil remediation technologies would be provided. Such an assessment – an evaluation of the social, ecological and economic pros and cons of technologies – contributes to a rational decision making process. The "soil" module did not yield such an assessment since government felt challenged by the researchers' initiative to inform the residents about soil pollution and since in this case there was no clearly identified addressee. The interaction of the academic and the bureaucratic policy cultures ended in a conflict about who had the authority to inform the residents. Such a conflict is what Gieryn calls boundary work (1983). Boundary work adjusts the boundary between the academic and the governmental policy culture, but it does not question the boundary as such. In the "soil" module transdisciplinary research aimed at going beyond informing by reorganizing knowledge. It was intended that the knowledge be used in a process of rational decision-making. In the end, however, boundary work was the only way the academic policy culture interacted with the other policy culture.

4.2. Type two: co-producing knowledge

Whereas type one projects remain close to the academic culture's policy to inform, type two projects facilitate a co-production of knowledge and at the same time participate in this process. Co-production means that the interaction between several policy cultures becomes a core element of the research process.

The "waste" project is a first example of type two. In this project, technologies for treating solid waste were developed in close collaboration with industry. Most of the projects were located at the research and development divisions of private companies. A natural scientist from a federal research institute was responsible for coordinating the module. He reported that the nature of the collaboration changed over time:

> First there were just self-interests. That was not easy to handle in the beginning. During the first meeting [...] we were all sitting around a table ... nobody dared to say anything, because everybody was afraid that the others would immediately spy on him or her. [...] To the end of the project people were rather open to each other in project matters, like colleagues. [...] And that was a piece of what I perceived as my work. That's a piece of – not sociology as a science – but practically used sociology, isn't it? (Natural scientist)

In order to make the competitors work in partnership, at least to some extent, collaboration and ownership had to be framed. Written rules, signed by the researchers, were established that prescribed when, and by whom, information could be classified as confidential; what that meant for the further handling of the information; and an agreed procedure for dealing with possible conflict. Besides this, the position of project observer was established. The observers came from industry, government or environmental organizations working in the field of waste. Again, they had to agree on certain rules and duties concerning information before being given the status of observer. The status of observer was established to anchor the project in the wider waste community.

> The point is, that the circle of those interested in waste management in our country is big, much bigger than the group of people involved in our module [...]. And that means: not only have sociologists, political scientists or economists to be involved, but also a much closer circle.[...] We have to include the people that do not form part of the module, but are closer to waste management than the outside world. (Natural scientist)

Rules of collaboration and the status of an observer were introduced to encourage researchers and observers to talk to each other, and to enhance mutual understanding, in order to make the module a collective endeavour. The collective nature of the collaboration was also addressed, when the coordinator reported on how he tried to orient the module's research towards sustainable development.

> I said to them: look, it is very clear to me that you are all potential competitors, must be like this, it's how our system works. But on the other hand it might be useful to have a club somewhere, that is not looking at the short-term profit only, but is also asking: "Where in fact is north?" [...] And I tried to tell them: "North is that the waste of today is the raw material of tomorrow". (Natural scientist)

In the "waste" module the researcher initiated an exchange of ideas within the module and with stakeholders of the waste management community. Furthermore, a normative orientation to the module's work was proposed. The desired direction ("north") would be to think long-term, not short. In doing so the researchers leave the academic culture's policy to inform and become participant facilitators of a collective process of co-production of knowledge. Not only the academic, but also the other policy cultures are framing the co-production. The economic policy culture is articulated by perceiving knowledge as something that can be owned, transformed to marketable products and sold. This is taken up in the project by way of the rules of ownership and by classifying knowledge. Moreover, the economic, the civic and the bureaucratic policy cultures are represented by the observers. The transdisciplinary researcher in such a co-production is less concerned with establishing and maintaining boundaries and more with defining procedural rules, enhancing mutual understanding, and proposing normative orientations to make the co-production a collective process of policy cultures.

The "coast" module provides a second example of type two. The module addressed the conflicting interests involved in using coastal zones:

> How in the coastal zone do we resolve the conflicts between different types of users of the environment? Some people want to use the coastal zone just to get rid of wastewater, others want to use it for bathing, others for fishing, and so on. And there will be conflicts. And how can these be solved? (Natural scientist)

Some of the stakeholders involved in that conflict – several fishing research institutes, in part affiliated to the fishery department – had their own projects in the module and had been in close contact with fishermen for many years. The researchers studying the conflict situation used these institutes as contacts to get in touch with the fishermen and to inform them about the study on conflict. The researchers judged the present handling of conflict as an indirect management that focused on technical questions of fishing rather than on the real issues. The researchers thus published conflict studies – one on farming mussels, and one on fishing –called outreach reports, written in a popularized language. The stakeholders eagerly requested these studies. It was intended that the outreach reports would make the conflict into an openly debated public issue. Asked about the role of research in the ongoing project and specifically in conflict resolution, one researcher stated:

> A problem may be that we are asked to produce what I would call technical conflict solutions, and to assess the pros and cons. What are the pros and cons if we decide in the conflict between the seals and fishing: "Hunting allowed for a defined period of time". You cannot say with all your scientific identity and persuasiveness: "Yes, that is the present state of knowledge." It is much more about reviving the stakeholders' attitude to look at things from different viewpoints. And that science has its limits, too.[…] And that – starting from the utopian model of transdisciplinarity – we have to develop solutions collectively and in conscious recognition that there is no monopoly for scientific knowledge. (Social scientist)

The researchers in the "coast" module induced a public debate on the conflict and its management. They informed the involved policy cultures by outreach reports consistent with the academic culture's policy to inform. Information, however, was primarily a means of initiating a public debate. The researchers saw their role as facilitators of a debate based, amongst other things, on scientific information; but without the academic policy culture being given the monopoly for informing the other cultures. The civic, the bureaucratic and the economic policy cultures were involved as stakeholders in the public debate. Again the boundaries and the way they could be upheld was of minor interest compared to question of how a collective search for conflict resolutions could be initiated.

EXTRACT 6.3

Leahey, E., Beckman, C. and Stanko, T. (2017) 'Prominent but less productive: The impact of interdisciplinarity on scientists' research', *Administrative Science Quarterly*, 62(1): 105–39.

Discussion

To empirically investigate the potential costs and the widely touted benefits of interdisciplinary scholarship, we collected and collated data from various sources for a sample of almost 900 center-based scientists and their 32,000 publications, providing the first systematic and mid-scale assessment of interdisciplinary research's impact. Our results demonstrate that IDR does benefit scientists: it improves their visibility in the scientific community, as indicated by cumulative citation counts. But we also document a productivity penalty associated with IDR: it depresses the number of articles that scientists publish. We see these effects in analyses of papers (i.e., high-IDR papers are more highly cited), in yearly analysis (e.g., scientists publishing high-IDR papers publish less in that year), and at the person level (e.g., scientists who publish IDR have more citations and higher variation in their citations). And when compared, the productivity penalty for IDR (standardized coefficient for direct effect = −.09) outweighs the reception benefit (standardized coefficient for direct effects = + .03) of engaging in this research. In other words, engaging in interdisciplinary research depresses productivity more than it increases citations. Compared with a scholar in the 20th percentile of IDR, a scholar in the 80th percentile of IDR produces seven fewer articles (20 rather than 27) and garners 40 more citations (230 rather than 190). The productivity penalty is strong enough to make the total effect of IDR on citations—i.e., the direct effects reported above, plus the indirect effect of IDR on citations via productivity—slightly negative (standardized coefficient for total effect = −.04). Apparently the learning curve is steep: it takes more time, effort, diligence, and perhaps coordination to master (at least aspects of) different fields and to work with scientists trained in disparate disciplines. That said, we cannot assess whether this trade-off is harmful or beneficial for a scientist—do 40 more citations offset seven fewer publications? We can say that the greater visibility this work receives is accompanied by fewer publications published.

We also make several theoretical contributions to the literature. First, we reorient away from reception penalties, which dominate the categories literature, toward production penalties, and we explore a new form of production penalty. The few recent papers that have examined production penalties focus on reduced quality (Kovács and Johnson, 2014), documenting through blind taste tests (i.e., controlling for audience perception), for example, that category-spanning wines are rated more negatively—presumably because of skill deficiencies on the part of the winemakers (Negro and Leung, 2013). In contrast, we focus on another form of production penalty that besets category-spanning work, at least in the realm of science: reduced productivity. Supplementary data on unpublished working papers, journal turnaround times, and a survey of authors' experiences, in addition to supplementary models incorporating individual fixed effects, suggest that productivity is hampered by the communication and coordination hurdles faced in the research process and not by the review process. This needs to be reconciled with the implicit production benefits found in the innovation and diversity literatures: better decisions (Beckman and Haunschild, 2002), heightened creativity (Pelled, Eisenhardt, and Xin, 1999), and organizational centrality (Powell, Koput, and Smith-Doerr, 1996). Our work suggests a potential short-run (fewer papers) and long-run (higher visibility) tradeoff.

Second, we demonstrate that production penalties and reception benefits depend on characteristics of the field. The production penalty (fewer papers) is less severe in a field that is increasingly interdisciplinary, like chemical engineering, than in a more stable field, whether that be a high-IDR field like life sciences or a low-IDR field like electrical engineering. The reception benefits, however, appear to accrue to high-IDR fields like life sciences. This suggests that the benefits and penalties of doing IDR work are not the same across fields. More broadly, the effect of category spanning depends on its typicality in the field (Lo and Kennedy, 2015). It may be that these high-IDR fields are shaped by high-status actors who reap the benefit of category spanning and normalize it for those who follow (Rao, Monin, and Durand, 2005; Sgourev and Althuizen, 2014).

Third, we move beyond mere category spanning to take into consideration the relationship between the spanned categories. Spanning two dissimilar entities is qualitatively different from spanning two similar entities, but extant research on both category spanning (Hsu, Hannan, and Kocak, 2009; Negro, Hannan, and Rao, 2010) and recombinant innovation (Fleming, Mingo, and Chen, 2007) largely ignores this (but see Leahey and Moody, 2014, and Kovács and Johnson, 2014). The IDR measure we use is sensitive to such differences, allowing connections between two unrelated disciplines to contribute more to the IDR score than connections between two related disciplines. The distinction is crucial and allows us to distinguish between the effects of variety—branching out into a number of other fields—and distance—branching out into cognitively distant and unrelated fields.

The distance measure we use is consistent with the logic of diversity and brokerage—which presumes networks with structural holes contain more distant, non-redundant knowledge—but we actually measure distance in knowledge space. When we measure IDR as sheer variety, neglecting distance between fields, we find that the reception benefit holds— papers that span dissimilar and even similar subfields receive more citations— but the productivity penalty does not, suggesting that the cognitive challenges and coordination costs associated with producing IDR are largely a function of the cognitive distance among fields. This is also the case when we calculate a measure of multidisciplinarity, which accounts for distance between fields represented in a scholar's oeuvre but not between fields represented in a given paper. Taken together, these results suggest that efforts to integrate, in a single paper, the cognitive distance across disparate fields contribute to the reception benefit (greater citations) and drive the productivity penalty (fewer publications).

Practically, should university research administrators and federal agencies like the National Academies of Science and the National Science Foundation continue to invest in IDR? Given that IDR scholars produce useful and noteworthy research that has more impact on the scientific community, the enthusiasm for IDR is not premature. Our analysis suggests, however, that it is not attuned to the implications—especially the negative implications—that IDR has for individuals. Scholars who produce interdisciplinary work may be more likely to publish in top-tier journals— the person-level correlation between IDR and journal impact factor is .25—but their overall productivity is hampered. There is a clear production penalty associated with interdisciplinary work. Even though our analysis of working papers suggests that IDR does not hamper subsequent publication, and if anything seems to help, additional analysis on a larger sample is warranted. The penalties of IDR may be more far-reaching than we document here.

Commentary

Sabine Hoffmann

I am writing this commentary as a tenured Group Leader for inter- and transdisciplinary research at the Swiss Federal Institute of Aquatic Science and Technology (Eawag) in Switzerland. In this position I currently perform a double role: as *leader* of the strategic inter- and transdisciplinary research programme, Water and sanitation innovations for non-grid solutions, WINGS, and as *core team member* of the two Swiss National Science Foundation (SNSF) Sinergia projects TRAPEGO (Transformation in Pesicide Governance) and TREBRIDGE (Transformation toward Resilient Ecosystems: Bridging natural and social sciences). I am strengthening communication, collaboration and integration across the boundaries not only of different disciplines (that is, interdisciplinarity), but also science, policy and practice (transdisciplinarity). As a *scholar* on inter- and transdisciplinary research, I am studying integration and integrative leadership in theory and practice using, among others, WINGS, TRAPEGO and TREBRIDGE as empirical cases. Wearing these two 'hats' enables me, on the one hand, to experience (in practice) the challenges and opportunities of enhancing communication, collaboration and integration in very different inter- and transdisciplinary contexts, and to use that experiential knowledge to develop new methodological and conceptual approaches. On the other hand, it enables me to study (from a theoretical standpoint) what constitutes (un)successful inter- and transdisciplinary research, and to apply that theoretical knowledge to address the communicative, collaborative and integrative challenges (and opportunities) inherent to inter- and transdisciplinary research. This commentary builds on these insights and embeds the three extracts in this chapter into more recent literature on knowledge production, dissemination and utilisation.

Inter- and transdisciplinary research is expected to benefit science and society (Extract 6.3). The underlying assumption of inter- and transdisciplinary research is that fruitful collaboration among scientific and societal actors, meaningful integration of different scientific and societal perspectives, and effective dissemination of inter- and transdisciplinary research findings to intended target groups enhances knowledge utilisation in the realms of both science and society, and ultimately fosters scientific and societal progress (see Hoffmann et al, 2019).

Christian Pohl's analysis (Extract 6.2) reveals two very different types of inter- and transdisciplinary research: the first type can be considered an interdisciplinary endeavour of researchers to reorganise knowledge from different disciplinary fields and to inform – imagined – target groups, thereby readjusting and even stabilising boundaries between science and society. The second type can be regarded as a transdisciplinary endeavour of researchers and stakeholders to co-produce knowledge and to disseminate that knowledge to intended target groups, thereby blurring or even dissolving boundaries between science and society. The second type of transdisciplinary research echoes recent conceptualisations of an ideal-typical, interactive and iterative transdisciplinary research process (Hoffmann et al, 2019), which acknowledge the 'fundamentally social ways in which knowledge emerges, circulates, and gets applied in practice' (Greenhalgh and Wieringa, 2011: 502). Such conceptualisations emphasise the need for more sustained and intense informal and formal interactions between researchers and intended target groups to ensure greater use of inter- and transdisciplinary research findings in science and practice at large, thereby dissolving boundaries between science and practice (Hoffmann et al, 2019).

In Extract 6.3, Erin Leahey et al analyse the professional costs and benefits associated with this second type of transdisciplinary endeavour. Their analysis shows that inter- and transdisciplinary research is a 'high-risk, high-reward endeavour' involving 'a potential short-run (fewer papers) and long-run (higher visibility) trade-off': engaging in such an endeavour increases, on the one hand, scientists' overall visibility in the scientific community (scientists publishing inter- and transdisciplinary papers have more citations, but higher variation of citations), but on the other hand, reduces their overall productivity (scientists publish less inter- and transdisciplinary papers in a year). Extract 6.3 mainly attributes this 'productivity penalty' to the cognitive and collaborative challenges associated with inter- and transdisciplinary research: Leahey et al highlight the time, effort, diligence and additional coordination required to work across disciplines. These findings resonate with Flinders et al (2016: 266), who argue that co-production

is time-consuming, ethically complex, emotionally demanding, inherently unstable, vulnerable to external shocks, subject to competing demands and expectations, and other scholars (journals, funders, and so on) may not even recognise its outputs as representing 'real' research. Academics who venture into the sphere of co-production are therefore not only likely to be stepping outside their own intellectual comfort zone, but they are also likely to be challenging dominant disciplinary norms and expectations.

Moreover, academics are often forced to meet traditional standards of disciplinary excellence *and* inter- and transdisciplinary excellence in order to attain promotion (Lyall, 2019) and earn a place in the current academic system, exposing them to a serious risk of physical and emotional exhaustion (Hoffmann et al, 2019).

Leahey et al's analysis in Extract 6.3 resonates with recent discussions on the challenges, costs and benefits of inter- and transdisciplinary research (see Oliver et al, 2019) and the multidimensional character of such challenges. Following Boix Mansilla et al (2015), Pohl et al (2021) distinguish at least three dimensions: (1) a *cognitive dimension*, which implies understanding distinct perspectives from different disciplines and fields, combining these perspectives and generating new knowledge by establishing previously unrecognised connections between them (Jahn et al, 2012; Specht et al, 2015); (2) a *social dimension*, which involves recognising, appreciating, balancing and, where possible, reconciling, different and sometimes competing interests, expectations and needs of involved team members, (sub)projects and organisations (Jahn et al, 2012: Hoffmann et al, 2017), and managing 'essential tensions' (Hackett, 2005), power imbalances and interpersonal conflicts inherent to inter- and transdisciplinary research (Oliver et al, 2019); and (3) an *emotional dimension*, which implies building a climate of trust, respect, recognition and appreciation among team members (Cronin et al, 2011; Boix Mansilla et al, 2015); it also involves providing a safe space that enables team members to learn from each other (and the challenges they face while engaging in inter- and transdisciplinary research) (Freeth and Caniglia, 2019).

Given the multidimensional character of these challenges, and the significant amount of time and resources this second type of transdisciplinary endeavour requires, it is no surprise that scientists engaging in such endeavours experience reduced productivity in terms of fewer publications per year. Based on their analysis, Leahey et al (Extract 6.3) therefore conclude that: 'future research should examine forms of productivity other than publishing, such as patenting, consulting, and advising, to examine whether scientists are compensating for their fewer publications with other activities.'

This conclusion echoes recent calls to counter the (unfortunately) persistent 'publish or perish' culture in academia, and to replace 'productivity' as the main quality criterion of research by 'societal relevance', 'credibility' and 'legitimacy' of co-produced research (see Durose et al, 2018; van den Besselaar et al, 2019).

To conclude, if academia is to be serious about fostering scientific and societal progress, it needs to recognise the need for more sustained and intense informal, and formal, interactions between scientific and societal actors in order to ensure greater use of inter- and transdisciplinary research findings in science and society at large. In practice, however, such interactions are still often based on one-way linear knowledge dissemination models that oversimplify such complex and multidirectional interactions (Sarkki et al, 2015). To ensure effective communication and productive collaboration across the boundaries of different disciplines, but also science, policy and practice – and, more importantly meaningful integration across such boundaries – academia needs to change its current incentive structures that (still) reward disciplinary science that does not engage with society in order to allow scientists with respective expertise in communication, collaboration and integration to thrive in the academic system (see also Hoffmann et al, 2022). Although such change constitutes a challenge, it is much needed if academia is to be serious about addressing the most pressing environmental and societal problems of our time.

Relating to my own role at Eawag as *leader* of and *scholar* on inter- and transdisciplinary research, I recognise that effective communication, not only across boundaries but also across hierarchies, is at the core of inter- and transdisciplinary research. It is a necessary prerequisite for productive collaboration and meaningful integration. In inter- and transdisciplinary teams I am involved in, effective communication is sometimes hampered, by, for instance, (1) negative forms of humour or sarcasm that break off discussions or devalue the validity of other perspectives; (2) manifest disinterest or disrespect that surface in behaviours such as arriving late at meetings, engaging in side conversations, checking emails or reviewing papers instead of engaging in team discussions; as well as (3) unproductive debates challenging the expertise of some, while (blindly) accepting the expertise of others (see also Thompson, 2009). Effective communication is facilitated, in contrast, by factors such as (1) spending time together and engaging in informal exchanges to become more familiar with each other; (2) creating a positive environment of mutual trust and respect to gain confidence about sharing individual perspectives; (3) recognising, reflecting and negotiating meanings of concepts and terms to transcend language differences, particularly between different disciplines; and (4) using positive forms of humour and shared laughter to reduce tensions and strengthen bonds among each other (see also Thompson, 2009). In my experience, working towards these factors positively influences a team's ability to communicate effectively to address complex environmental and societal problems in a collaborative and integrative way.

References and further reading

Boix Mansilla, V., Lamont, M. and Sato, K. (2015) 'Shared cognitive-emotional-interactional platforms: Markers and conditions for successful interdisciplinary collaborations', *Science, Technology & Human Values*, 41(4): 571–612, https://doi.org/10.1177/0162243915614103

Cronin, M.A., Bezrukova, K., Weingart, L.R. and Tinsley, C.H. (2011) 'Subgroups within a team: The role of cognitive and affective integration', *Journal of Organizational Behavior*, 32(6): 831–49, https://doi.org/10.1002/job.707

Durose, C., Richardson, L. and Perry, B. (2018) 'Craft metrics to value co-production', *Nature*, 562: 32–3, DOI:10.1038/d41586-018-06860-w.

Flinders, M., Wood, M. and Cunningham, M. (2016) 'The politics of co-production: Risks, limits and pollution', *Evidence & Policy*, 12(2): 261–79, DOI:10.1332/174426415X14412037949967.

Freeth, R. and Caniglia, G. (2019) 'Learning to collaborate while collaborating: Advancing interdisciplinary sustainability research', *Sustainability Science*, 15: 247–61, DOI:10.1007/s11625-019-00701-z.

Greenhalgh, T. and Wieringa, S. (2011) 'Is it time to drop the "knowledge translation" metaphor? A critical literature review', *Journal of the Royal Society of Medicine*, 104(12): 501–9, https://doi.org/10.1258/jrsm.2011.110285

Hackett, E.J. (2005) 'Essential tensions: Identity, control, and risk in research', *Social Studies of Science*, 35(5): 787–826, https://doi.org/10.1177/0306312705056045

Hoffmann, S., Klein, J.T. and Pohl, C. (2019) 'Linking transdisciplinary research projects with science and practice at large: Introducing insights from knowledge utilization', *Environmental Science & Policy*, 102: 36–42, www.dora.lib4ri.ch/eawag/islandora/object/eawag:19310

Hoffmann, S., Pohl, C. and Hering, J.G. (2017) 'Exploring transdisciplinary integration within a large research program: Empirical lessons from four thematic synthesis processes', *Research Policy*, 46(3): 678–92, www.dora.lib4ri.ch/eawag/islandora/object/eawag:13859

Hoffmann, S., Deutsch, L., Klein, J.T. and O'Rourke, M. (2022) 'Integrate the integrators! A call for establishing academic careers for integration experts', *Humanities and Social Sciences Communications*, 9: 147, https://doi.org/10.1057/s41599-022-01138-z

Holbrook, B.J. (2017) 'Peer Review, Interdisciplinarity, and Serendipity', in R. Frodeman (ed) *The Oxford Handbook of Interdisciplinarity*, Oxford: Oxford University Press, Chapter 34.

Jahn, T., Bergmann, M. and Keil, F. (2012) 'Transdisciplinarity: Between mainstreaming and marginalization', *Ecological Economics*, 79(July): 1–10, https://doi.org/10.1016/j.ecolecon.2012.04.017

Kukkonen, T. and Cooper, A. (2017) 'An arts-based knowledge translation (ABKT) planning framework for researchers', *Evidence and Policy*, 15(2): 293–311, https://doi.org/10.1332/174426417X15006249072134

Lyall, C. (2019) *Being an Interdisciplinary Academic: How Institutions Shape University Careers*, Basingstoke: Palgrave Pivot.

Oliver, K., Kothari, A. and Mays, N. (2019) 'The dark side of coproduction: Do the costs outweigh the benefits for health research?', *Health Research Policy and Systems*, 17(1): article 33, DOI:10.1186/s12961-019-0432-3.

Pohl, C., Klein, J.T., Hoffmann, S., Mitchell, C. and Fam, D. (2021) 'Conceptualising transdisciplinary integration as a multidimensional interactive process', *Environmental Science & Policy*, 118: 18–26, https://doi.org/10.3929/ethz-b-000467221

Pohl, C., Wülser, G., Bebi, P., Bugmann, H., Buttler, A., Elkin, C., Grêt-Regamey, A., Hirschi, C., Le, Q.B., Peringer, A., Rigling, A., Seidl, R. and Huber, R. (2015) 'How to successfully publish interdisciplinary research: Learning from an *Ecology and Society* Special Feature', *Ecology and Society*, 20(2): article 23, http://dx.doi.org/10.5751/ES-07448-200223

Sarkki, S., Tinch, R., Niemelä, J., Heink, U., Waylen, K., TIimaeus, J., Young, J., Watt, A., Neßhöver, C. and van den Hove, S. (2015) 'Adding "iterativity" to the credibility, relevance, legitimacy: A novel scheme to highlight dynamic aspects of science–policy interfaces', *Environmental Science & Policy*, 54: 505–12, DOI:10.1016/j.envsci.2015.02.016.

Specht, A., Gordon, I.J., Groves, R.H., Lambers, H. and Phinn, S.R. (2015) 'Catalysing transdisciplinary synthesis in ecosystem science and management', *Science of the Total Environment*, 534: 1–3, DOI:10.1016/j.scitotenv.2015.06.044.

Thompson, J.L. (2009) 'Building collective communication competence in interdisciplinary research teams', *Journal of Applied Communication Research*, 37(3): 278–97, https://doi.org/10.1080/00909880903025911

van den Besselaar, P., Sandstrom, U., Howaldt, J. and De Rijcke, S. (2019) 'Pathways to Impact from SSH research', *Journal for Research and Technology Policy Evaluation*, 48 (Proceedings of the Conference 'Impact of Social Sciences and Humanities for a European Research Agenda Valuation of SSH in mission-oriented research').

7

Improving Research Skills

Chapter overview

If you are based in a research-focused university, you will probably benefit from institutional research support and advice to help you conduct your research and, as a researcher, you will most likely already be equipped with the skills needed to perform research within your discipline. However, doing inter- and transdisciplinary research successfully requires additional 'meta-skills', as alluded to in the selected readings in this chapter. These meta-skills encompass such attributes as leadership, communication, negotiation and integration, and help us to understand why research across disciplinary cultures can sometimes be so challenging. Aspects of leadership are further discussed in Chapter 8 in this book.

The first extract from Julie Thompson Klein (Extract 7.1) comes from an edited collection that provides essential tools for individuals or teams developing an inter- or transdisciplinary project. It includes theoretical perspectives, case studies, communication tools and reports from different institutional contexts, all focusing on how to improve communication in collaborative settings. Other resources can be found in each chapter that complement the strategies presented in the book.

Erin Leahey's extract (Extract 7.2) reviews trends in the practice and study of research collaboration. Using as its source evidence from journal publications, Leahey classifies different types of collaboration, and analyses the drivers and consequences of collaboration for both individual researchers and for science more generally. Examples include the impact of collaboration on the contributing authors and their work, the use of multiple methods and measures, and research integrity. Readers might then consider these perspectives when assessing their own skills as 'integrative' researchers and practitioners.

If Leahey draws her evidence from quantitative analysis of publication trends, then the last extract in this chapter, from Myra Strober (Extract 7.3), offers a more qualitative approach to understanding core interdisciplinary skills and traits. Despite a willingness to engage in interdisciplinary conversations, scholars can experience difficulties shrugging off their discipline–based straitjackets when attempting to bridge disciplinary cultures. Strober observed a series of cross-discipline seminars within universities in the USA, and conducted follow-up interviews with participants to better understand how the habits and skills we learn as part of our discipline-based training can unconsciously affect our abilities to interact with scholars from other academic traditions.

EXTRACT 7.1

Klein J.T. (2013) 'Communication and Collaboration in Interdisciplinary Research', in M. O'Rourke, S. Crowley, S. Eigenbrode and J.D. Wulfhorst (eds) *Enhancing Communication and Collaboration in Interdisciplinary Research*, Thousand Oaks, CA: SAGE, Chapter 2.

[Note: In this extract, Klein uses the acronym C^2 to denote 'communication and collaboration'.]

Management

Studies of IDR in the management literature tend to focus on organizing teams and facilitating interactions. The formal study of teams, James Davis (1995) recalled in his book on interdisciplinary team teaching, started in employment settings. Early studies of group behavior, Davis added, evolved into more specialized studies of human communication and the social psychology of groups. Interdisciplinary task force management has also been a feature of military operations, civilian affairs, engineering projects, feasibility studies, and industrial research and development. World War II was a watershed in collaborative IDR, highlighted by major interdisciplinary initiatives, including the Manhattan Project and problem-focused operations research. Communication researchers also cite the World War II era as a heyday for small group communications, preceded by studies in the 1920s that adapted John Dewey's work on democratic group decision making. By the 1970s and 1980s, international competition in science-based industries heightened the demand for collaborative IDR, especially in manufacturing, computer sciences, biomedicine and pharmaceuticals, and high technology. This development, in turn, prompted new studies of large-scale complex projects, including the Human Genome Initiative (Klein, 1996, pp. 173–208; 2010a, pp. 16–21).

The early literature tended to apply management and organizational theories of the day to studies of IDR collaboration, with emphasis on organizational structures, leadership strategies, and types of teams. Over time, the focus expanded from managing teams and organizational units to creating institutional research cultures and the behavioral dynamics of collaboration. In a study of leadership in TDR, Barbara Gray

(2008) identified three general categories of tasks for enhancing collaboration: cognitive, structural, and processual. The success of C^2 depends in no small part on management of structural components that require coordinating tasks, internal linkages, and information flows. The organizational chart and task distribution must allow time for interaction, joint activities, coordinated use of common research facilities and instruments, consensus building, shared decision making, and networking across subprojects. Thompson (2009) came to a similar conclusion in urging regular attention to CCC, and Hindenlang, Heeb, and Roux (2008) highlight the role of platforms for handling structural tasks. Operating as loosely structured social networks, platforms create a space for communication, fostering mutual understanding, shared goals, concrete ideas and measures, and common assessment.

Even with strong platforms in place, conflicts arise. Status conflicts are especially tenacious. They arise for many reasons, including disciplinary and professional pecking orders as well as differences between quantitative versus qualitative approaches, academic rank, and gender, race, and cultural background. The theory of "status concordance" holds that success is linked to matched and equal factors. Rarely, though, do perfect matches occur (Klein, 1990b, pp. 127–128). Members of IDR teams also exhibit many of the same fighting and thwarting behaviors as other groups, echoing Bruce Tuckman's (1965) model of group development in stages of forming, storming, norming, performing, and (a fifth stage added later) advocating or transforming (see also Tuckman & Jensen, 1977). Conflict is associated with both technical issues (definition of a problem, methodologies, and scheduling) and interpersonal issues (leadership style and disciplinary ethnocentrism). Interdisciplinary teams must also overcome "boundaries of reticence" that disciplinary socialization creates, to avoid defaulting to disciplinary worldviews. Individuals must also grant power to others and surrender some degree of control (Caudill & Roberts, 1951; Stone, 1969).

Difference, though, can be an asset. The consent/dissent (Alteritaet) structure necessary for all communication, Vosskamp (1994) advises, shapes the possibility of interdisciplinary dialogue. Misunderstandings, animosities, and competitions must be taken seriously, not mitigated or glossed over. Moreover, even if differences are negotiated and mediated, they do not go away. They recur as participants work through their differences and attempt to resolve them in the interest of a common goal. In her pioneering study of working relationships among specialists in mental-health projects, Margaret Barron Luszki (1958) reported that members of interdisciplinary teams paid a price for congeniality in the short run. By not dealing with conflicts in definitions of core terms, such as aggression, they reduced the number of creative problem-solving conflicts that would have promoted high-level shared concepts in the long run. Certain ideal characteristics of interdisciplinary individuals have been identified, including flexibility, patience, willingness to learn, sensitivity toward and tolerance of others, reliability, and openness to diversity, new roles, and risk. These are ideals, however (Klein, 1990b, p. 183; Strober, 2011, p. 121). Participants are usually unwilling to abstain from approaching a topic from their own worldviews. Yet, Bruce Thiessen (1998, pp. 49–50) admonishes, adaptive behavior is required to achieve common ground in both language and goal directedness.

The *Collaboration and Team Science* field guide from the Office of the Ombudsman at NIH (Bennett & Gadlin, Chap. 17 this volume) emphasizes the importance of managing tensions. One of the recommendations for communicating about science is to establish ground rules for how participants will be expected to communicate in meetings. The distinction between dialogue and debate lies at the heart of the ground rules. Dialogue is a collaborative act of working together toward common understanding, rather than being oppositional. Common ground is the goal, not being close-minded, winning individual points, or defending one's position as the "best." An open-minded attitude and openness to being wrong and to change are needed. Listening to others provides a basis for agreement or consensus, along with seeing strengths in others' positions rather than flaws and weaknesses (Bennett et al., 2010, pp. 29–30).

The insights gained from findings in pertinent literatures come together in the overriding thematics of integration and learning in interdisciplinary C^2.

Integration

Integration is often regarded as the leading candidate for a distinguishing characteristic of collaborative IDR and its baseline vocabulary (see Holbrook, 2012; Repko, 2010). However, there is no universal theory or model of integration because the scope, degree of coordination of perspectives, nature of interactions, and goals of projects and programs vary too widely to allow a single coherent theoretical perspective. For example, teams organizations and nongovernmental organizations. Yet Pohl, van Kerkhoff, Hirsch, and Bammer (2008) identified four primary classes of "tools" for integration: (1) mutual understanding through communication, (2) theoretical concepts, (3) models, and (4) methods. This last category includes some well-known methods that have been widely applied, such as systems theory and modeling, Delphi and scenario building, simulation, concept mapping, computer synthesis of data and information flow, and integrated environmental assessment and risk management. Other methods target communication processes, including facilitating common understanding, mental mapping of stakeholder views, consensus conferences, and collaborative learning.

Yet, integration in collaborative IDR is a social process that requires coconstruction of common ground, even when using well-known approaches. Optimal integration, Davis (1995) exhorts, requires high levels of collaboration: "The greater the level of integration desired, the higher the level of collaboration required" (p. 20). Joint definition of a project is required, along with the core research problem, questions, and goals. Role clarification and negotiation help members assess what they need and expect from each other while clarifying differences in disciplinary language and approaches. And, ongoing communication and interaction foster mutual learning and interdependence, expanding individual identities into group identity. Young teams, Stone (1969) found, exhibit secondary-group relations. Members are self-protective, thinking in terms of "I." Primary-group relations are characterized by dedication to a common task, thinking in terms of "we." When the singularity of individual disciplinary identity is called into question, Holbrook suggests, Bataille's notion of strong interdisciplinary communication is operative.

In an analogy that will ring true for veterans of collaborative research, Koepp-Baker (1979) likens an interdisciplinary health care team to a polygamous marriage. The team is launched by announcement of intentions, engagement, publicity, a honeymoon, and finally the long haul (Koepp-Baker, 1979, p. 54). The Collaborators Pre-Nup, adapted from NIH, highlights the importance of early discussion of goals, roles, coauthorship, ownership of intellectual property, and obligations of teamwork (Ledford, 2008). From the outset, contextual factors also influence "collaborative readiness." In their studies of TDR centers, Stokols et al. (2010; Stokols, Misra, Moser, Hall, & Taylor, 2008) concluded, the more contextual factors in place at the outset, the greater the chances of success. Antecedents span interpersonal, environmental, and organizational parameters of research:

- Institutional supports for collaboration
- Breadth of disciplines, departments, and institutions at a center
- Prior experience working as team members on projects
- Spatial proximity or distance of offices and laboratories
- Electronic linkages (Stokols et al., 2008, 2010)

Making it through the "long haul" requires ongoing management of integrative process. Processual tasks, Gray (2008) stipulates, ensure constructive and productive interactions, with subtasks devoted to designing meetings, determining ground rules, identifying tasks that move partners toward common objectives, building trust, and ensuring effective communication. Defila et al.'s (2006) literature review of case studies identified more than 500 tips. And, McDonald, Bammer, and Deane's (2009) repertoire of dialogue methods includes hypothesis and model building, integrative assessment procedures, boundary objects and concepts, heuristics, research questions, artifacts and products, mutual learning, and stakeholder participation. Defila et al. (2006) also consider recursive procedure to be a general design principle of TDR. Iterative peer editing is one of the most common methods, fostering coassessment of individual contributions, collective reconciliation of differences, and greater likelihood of moving beyond multidisciplinary juxtaposition to interdisciplinary integration. Ideal models, Maurice DeWachter (1982) counsels, start with the assumption that individuals will suspend their disciplinary/professional worldviews. Yet his experience in bioethical decision making indicates the best chance of success lies in starting by translating a global question into the specific language of each discipline then working back and forth in iterative fashion. By constantly checking the relevance of each answer to a core question, no single answer is privileged. This process clearly entails the second overriding theme—learning.

Learning

Even with the current heightened profile of IDR, educational needs are underserved. The top recommendations for students in the National Academy of Sciences report on *Facilitating Interdisciplinarity Research* include taking a broad range of courses while developing a solid background in one discipline (Committee on Facilitating Interdisciplinary Research, 2004).

Undergraduates in particular are urged to seek courses at the interfaces of traditional disciplines that address basic research problems, courses studying social problems, research experiences spanning more than one discipline, and opportunities to work with faculty with expertise in both their disciplines and interdisciplinary process. Graduate students are encouraged to gain knowledge and skills in one or more fields beyond their primary area: by doing theses or dissertations with advisers from different disciplines, participating in conferences outside their primary fields, and, for all students, working with mentors from more than one discipline. The top recommendations for educators are to develop curricula incorporating interdisciplinary concepts, offer more interdisciplinary studies courses, take teacher-development courses on interdisciplinary topics and methods of teaching nonmajors, and provide opportunities that relate foundation courses, data gathering and analysis, and research activities to other fields of study and to society at large. The report also urged more training across the board in interdisciplinary research techniques, team management skills, and summer immersion experiences for learning new disciplinary languages and cultures (Committee on Facilitating Interdisciplinary Research, 2004).

Professional organizations have also called for greater attention to the need for interdisciplinary education. Pellmar and Eisenberg's (2000) *Bridging Disciplines in the Brain, Behavioral, and Clinical Sciences* presents models for training at all levels, from undergraduate curricula through postdoctoral fellowships, predoctoral and postdoctoral training programs, and career-long opportunities. The targeted areas for improving communication in particular are jargon, intellectual turf, team building, leadership, and interactions within physical spaces. *BIO 2010* (National Research Council, 2003) presents a blueprint for aligning undergraduate education in biology with contemporary research in a curriculum that integrates physical sciences with information technology and mathematics with life sciences. Both reports also expand the locus of learning from traditional academic curricula to open spaces of faculty development programs and training modules in situ. Training for IDR and TDR shares many features with traditional programs, but its distinct emphasis, Justin Nash (2008) points out, is developing researchers who can synthesize theoretical and methodological approaches from multiple disciplines. The most common form of training is a multimentor apprenticeship model, with mentors in separate disciplines in addition to multiple faculty advisors and time in residence in centers (Nash, 2008, p. S133).

The principle of mutual learning also repeats across levels, from the pragmatics of daily work to theory and epistemology. "Mutual learning" requires knowing how to recognize ignorance of a particular area, then soliciting and gathering new information and knowledge. In defining transdisciplinarity in domains of environmental literacy, Roland Scholz (2011) stresses processes of capacity/competence building, consensus building, analytic mediation, and legitimation of public policy. Generating "socially robust knowledge" when working with stakeholders in public and private sectors goes beyond listening to their inputs to actually engaging them in the research process in a manner that provides feedback to the generation of scientific knowledge and theory building (Scholz, 2011, pp. 373–374). Knowledge is not simply exchanged. It is constructed as individuals with differing views and stakes work together.

In defining the three major forms of collaboration in TDR—common group learning, deliberation among experts, and integration by a subgroup or individual—Pohl et al. (2008) capture the principle of mutual learning at the heart of the first form. "Common group learning," they admonish, "means that integration takes place as a learning process of the whole group" (p. 415). Management is still crucial, however. Leaders need to coach the process by promoting joint learning activities. Adopting a conscious, targeted approach to communication, Schmithals and Berhenhage also urge, is crucial for integrating knowledge and methods when working across cognitive cultures (cited in Bergmann et al., 2010). A project-specific "cooperation and communication culture" must be established, with attention paid to interfaces: to points where the work of one participant is necessary for the work of another and to points where participants must coordinate effectively (Schmithals & Berhenhage, cited in Bergmann et al., 2010). For leaders, creating a common culture also entails the third of Gray's (2008) general categories of tasks for enhancing collaboration—cognition. The cognitive task requires managing sensemaking by creating a mental model, map, or mind-set of goals and ways of achieving them through visioning and framing (Gray, 2008, pp. S125–S126).

In a case study of an interdisciplinary team focused on STEM education (science, technology, engineering, and mathematics), DuRussel and Derry (2011) integrate conceptual understandings of mental models with the concept of *situation awareness* in organizational literature. Social and cognitive integration, DuRussel and Derry affirm, are tightly interwoven in situation awareness. An individual's prior knowledge helps shape his or her mental model (i.e., cognitive representation) of a particular situation. Yet models are dynamic and changing. As situational learning occurs, new models influence and change the content and structure of permanent schemas. The key insight is *alignment* of mental models. Common features correlate positively with productivity, but, akin to the Toolbox Project, models must be made explicit if common understanding is to be achieved. Contexts range from goals and objectives of projects to the variety of tools that mediate between individual work and team objectives. They also include *norms* and *rules* that influence communication patterns, as well as distribution of tasks and patterns of interactions, including regular participation in meetings and common criteria of evaluation. Visioning and (re)framing models stimulate ideas about how disciplines might overlap in constructive ways that generate new understandings and encourage collaborative work modes.

In a project that brings together learning theory, epistemology, and sociology of knowledge, Boix Mansilla, Lamont, and Sato (forthcoming-a, forthcoming-b) posit an emerging model of *sociocognitive platforms for interdisciplinary collaboration*. The team studied experts working in nine established research programs and networks. Their case studies yield an empirically based picture of the *social-interactive* ways participants construct group membership and collective norms, and the ways norms contribute to or hinder creation of common platforms. They also considered the *cognitive-epistemic* ways experts define their enterprise and platforms for integration. Preliminary data from five networks affirm the need for early investments of time and effort to define a problem space and approach. The group climate of networks differs, though, influencing problem definition, intellectual agendas, degrees of exchange and integration, and patterns of dominance in particular disciplinary mixes. One network, for instance, had a just-in-time approach. Its "agile" and "opportunistic" work style contrasted with the prescriptive style of another group. Contrasts in emotional intensity also appeared, and products that function as boundary objects varied. Products provide "concrete space" for interaction, coproduction, and disciplinary translation in the form, for example, of graphics, concept maps, and constructs. Platforms also change over time. Gains in communicative and collaborative capacity in particular include greater clarity about disciplinary languages, increased comfort with unknown terrains, and recalibrated beliefs about another discipline. Microsocial networks realign, too, with growing "deliberative competency" at the group level and individual-level sociocognitive gains, such as the ability to provide "honest and constructive feedback" (Boix Mansilla et al., forthcoming-a, forthcoming-b).

Conclusion

A final example bridges the two core concepts engaged in this investigation—communication and collaboration. Vosskamp (1994) and Klein (1996) treat interdisciplinarity as communicative action, and in a major study of an urban planning project, Després and colleagues (2008) extended this idea. Scientific and academic knowledge alone, they argue, cannot deal adequately with the complexity of subjects and problem domains such as revitalization of neighborhoods, including their case study of retrofitting older residential neighborhoods on the outskirts of Quebec City. Following Habermas's (1987) *Theory of Communicative Action*, instrumental, ethical, and aesthetic forms of knowledge are all needed. Moreover, rational knowledge comes out of not only "what we know" but "how we communicate" it, generating a form of "communicative rationality." Stakeholders entering into negotiation confront the four kinds of knowledge in a series of encounters that allow representatives of each type to express their views and proposals. In the process, a fifth type of knowledge progressively emerges. It is a hybrid product, the result of "making sense together."

Fostering "intersubjectivity," the fifth type of knowledge requires ongoing efforts to achieve mutual understanding, aided in this case by a mediator who helped extract individual interests or views. As progressively shared meanings, diagnoses, and objectives emerge, individual interests and views are seen in different perspectives.

The underlying premises of this case study—communicative action, communicative rationality, negotiation, intersubjectivity, and mutual understanding—bring us full circle back to philosophy. Anne Balsamo (2011, p. 163; see also Balsamo & Mitcham, 2010, p. 270) frames that return in ethical principles of collaboration. The first two principles accentuate requirements for individuals:

> *Intellectual confidence*: The understanding that one has something important to contribute to the collaborative process. This is the commitment that makes one accountable for the quality of an individual's contribution to the collaboration. Everyone's contribution to the collaboration must be reliable. It must be thorough and full of integrity; it must refuse shortcuts and guard against intellectual laziness.

Intellectual humility: The understanding that one's knowledge is always partial and incomplete and can always be extended and revised by insights from others. This is the quality that allows people to admit they don't know something without suffering loss of confidence or a blow to their self-esteem.

The second two principles move from individuality to group responsibility:

Intellectual generosity: The sincere acknowledgment of the work of others. This acknowledgment must be explicitly expressed to collaborators as well as through citation practices. Showing appreciation for other ideas in face-to-face dialogue and throughout the process of collaborative process sows the seeds for intellectual risk taking and courageous acts of creativity.

Intellectual flexibility: The ability to change one's perspective based on new insights that come from other people. This is the capacity both for play and reimagining the rules of reality: to suspend judgment and to imagine other ways of being in the world, and other worlds to be within.

An overriding principle of integrity emerges from the move to group responsibility:

Intellectual integrity: The habit of responsible participation that serves as a basis for the development of trust among collaborators. This is a quality that compels colleagues to bring their best work and contribute their best thinking to collaborative efforts.

Skills of research integration, McDonald et al. (2009) assert, have become as essential today as disciplinary skills. These skills are all the more important when disciplines are undergoing changes characterized by greater boundary crossing, openness to interdisciplinary developments, prioritizing of real-world problems, and grappling with the complexities of contingency, contextualization, and diversity. However, much of the wisdom of practice is not captured for future use. Handbooks and networks are vital forums for disseminating wisdom of practice, highlighted recently by a new online suite of team science learning modules on the Team Science website (http://teamscience.net/) and NIH Team Science Toolkit (http://www.teamsciencetoolkit.cancer.gov/public/Home.aspx). However, "expert praxis" does not lie simply in formulas, well-honed guidelines, or tested models. It is emergent from communicative actions among the participants. O'Donnell and Derry (2005) liken the challenge that interdisciplinary teams face to Krauss and Fussell's (1990) concept of the "'mutual knowledge' problem." Experts within a discipline typically share a "common referential base," aiding communication. Participants in collaborative IDR must develop "a shared knowledge base" that constitutes "group intelligence" (O'Donnell & Derry, 2005, pp. 73, 76-77).

EXTRACT 7.2

Leahey, E. (2016) 'From sole investigator to team scientist: Trends in the practice and study of research collaboration', *Annual Review of Sociology*, 42(1): 81–100.

Benefits of collaboration

Empirical work on collaboration focuses heavily on its beneficial outcomes, possibly because the perceived and real benefits of collaboration contributed to its rise (Wray 2002). There is evidence that collaboration increases productivity. Rawlings and McFarland (2011) find that collaborating increases grant activity (number of submissions as well as number of awards and dollar amount). In terms of publishing, Lee & Bozeman (2005) find that collaborating has a positive effect on a "normal count" of publications (adding more lines to one's CV) but has no effect on a "fractional count" (in which the number of publications is divided by the number of authors). Even if contributing authors do not gain more fractional publications from collaborating (Bikard et al. 2015), they do attain greater visibility on average. Collaborative teams produce more high-impact articles and garner more citations (Fox 1991, Lee & Bozeman 2005, Shi et al. 2009). Wuchty and colleagues (2007, p. 1037) find a "broad tendency for teams to produce more highly cited work than individual authors." Although this could be attributed simply to the greater number of coauthors who can share the work with their diverse sets of contacts (Bikard et al. 2015, p. 7), it could also be attributed to the higher quality of the work that results when people collaborate, bounce ideas off each other, and check each other's work. As Abramo and colleagues (2014, p. 2290) find, "collaborations permit participation in broader research projects, access to funding, and not least, improvement in personal competencies, with positive effects on the quantity and quality of publications." Some scholars have noted that collaboration brings legitimacy to an idea (Bikard et al. 2015); presumably, if more than one scientist believes enough in a project to pursue it, then that project is more worthwhile than sole-authored work that does not have such obvious backing. Although the benefits of collaboration vary by discipline (Walsh & Bayma 1996), in general collaboration ensures funding flows and career advancement (Abramo et al. 2014, Jeong & Choi 2015). Clearly, collaboration is beneficial for individual scientists (Presser 1980) and perhaps for scientific progress more generally (Hara et al. 2003).

Costs of collaboration

One of the more promising and sociologically minded developments in this literature is the increasing recognition (and investigation) of the costs of collaboration, which begins to balance the heavy focus on benefits. Cummings & Kiesler (2005, 2007) and Shrum and colleagues (2007) have documented the communication and coordination costs that accompany collaboration, especially collaborations that cross disciplinary and/or institutional bounds. Often, the goals of team members are not aligned, communication is difficult because team members lack a common language and set of experiences, and translation is needed. Moreover, processes and routines need to be developed so that work can be distributed and synchronized and progress can be monitored effectively (Bikard et al. 2015). Such articulation work (Fujimura 1996)—managing people, maintaining hardware and software and material supplies, communicating with coauthors, and updating funding agencies via reports—falls on the head of a laboratory scientist and draws him or her away from the actual research (Hackett 2005). It is possible that these additional administrative tasks compromise real productivity. As noted earlier, collaborating allows one to publish more articles (so there will be more lines on one's CV; see Crane 1972, Lee & Bozeman 2005), but fractional productivity (toward which sole-authored articles contribute 1, articles with two authors contribute 0.5, articles with three authors contribute 0.33, etc.) is not bolstered and indeed may decline (Bikard et al. 2015, Lee & Bozeman 2005). Thus, the "coordination difficulties stemming from collaboration in creative work are generally associated with a loss of individual productivity" (Bikard et al. 2015, p. 6). Presumably, because (at least for promotion and tenure purposes) all citations of an article are credited to all coauthors, reduced productivity is offset by enhanced visibility (Bikard et al. 2015).

A few scholars in particular have conducted in-depth investigations of the potential costs and tensions inherent in collaboration by conducting interviews with team leaders and members. Boardman & Bozeman (2007) highlight the role strain that besets scientists engaging in collaboration via research centers. These individuals are subject to competing demands in the workplace, and this can take a toll on their job satisfaction, productivity, and emotional well-being. Whereas these competing demands results largely from distinct organizational allegiances (e.g., to one's department or research center) and from the centers' foci on interdisciplinary work and technology transfer (both of which require domain spanning), it is likely that role strain accompanies scholars in more general types of collaborations, too, as they figure out how to balance an independent research program with collaborative work, as well as in multiple collaborations. Hackett (2005) identifies this tension between independence and dependence and also explores other polarities, including cooperation/competition, openness/secrecy, democracy/autocracy, and risk/control in his study of molecular biology labs. Through an in-depth case study of one collaborative team in ecology, Parker & Hackett (2012, p. 21) identify the emotional work that, while enhancing creativity, also make skepticism "more likely to occur and more challenging to manage." Initial collaborative success prompts growth in the size and diversity of the team, but these changes endanger the conditions that prompted success in the first place, producing yet more tensions in collaborative work. McBee & Leahey (2016) identify numerous challenges faced by even the most successful interdisciplinary humanities scholars, including a lack of formal and informal support, dampened productivity, and devaluation in the peer-review process.

Another cost of collaboration is free riding: Some coauthors do not do much (or any) of the work but are still included as coauthors. How common are guest authors whose names appear in the list of authors, but who in reality contributed very little to the article? We do not know; the evidence is largely anecdotal. Empirical research—probably at the team level, and probably qualitative—is needed to ascertain the prevalence of this practice in scientific research. In an effort to stymie the inclusion of guest authors, some top journals (e.g., Nature, Science, PNAS) have encouraged authors to delineate how each author contributed to the article. However, this practice remains rare. It would be interesting to track the diffusion of this practice across journals and disciplines; according to Endersby (1996, p. 390), "the ideal standard for attribution of credit may be to footnote each author's responsibilities and percentage of credit for published research." Regardless of individual practice and emerging journal policy, the American Sociological Association Code of Ethics stipulates that: (a) "Sociologists take responsibility and credit, including authorship credit, only for work they have actually performed or to which they have contributed"; and (b) "Sociologists ensure that principal authorship and other publication credits are based on the relative scientific or professional contributions of the individuals involved, regardless of their status. In claiming or determining the ordering of authorship, sociologists seek to reflect accurately the contributions of main participants in the research and writing process" (see http://www.asanet.org/images/asa/docs/pdf/CodeofEthics.pdf).

Another underdeveloped concern with collaboration is potential exploitation. At least two studies of types of collaboration have found evidence of advisor-advisee collaborations, which "raise many particularly sensitive issues" (Endersby 1996, p. 385). Leahey & Reikowsky (2008) identify a "mentoring" style of collaboration, wherein established scholars join with scholars-in-training to publish research together; this style characterized about 10% of the sociology articles they sampled. Their supplementary interview data suggest that this collaborative style is on the rise due to job market and tenure expectations for productivity. As one interviewee stated, "We've become more generous as a discipline and we're more willing to give graduate students the opportunity to collaborate" (Leahey & Reikowsky 2008, p. 435). Bozeman & Corley's (2004) principal components analysis of survey data from a sample of 1,000 scientists also found a mentoring style of collaboration to be common among scholars who desire to help junior colleagues and graduate students. They do not provide a sense of how common this style is, but they do indicate that mentors are more open to industry collaboration and tend to collaborate with graduate students and junior colleagues who are women. Collaborating with students and junior colleagues in publishing is one of the most helpful things an advisor can do (Long & McGinness 1981), and research documents that women are not only more likely to collaborate than men (Abramo et al. 2013, 2014), but they are also more likely to benefit, in terms of productivity, from it (Kyvik & Teigen 1996). Hu and colleagues (2014) find that senior faculty benefit more from collaborating than junior faculty. Thus, it appears that the potential for exploitation lies not in women's and younger scholars' contributions being neglected completely (indeed, they are listed as authors), but in the fact that they might not receive appropriate credit [i.e., they do most of the work but are listed only as contributing, not lead, authors (Larivière et al. 2015b)]; and even if they are the lead authors, audiences may associate the article with the senior author's name, to the detriment of the junior scholars' visibility.

This begs the important question of credit allocation. With increasing collaboration, new approaches to evaluation will be necessary. This is especially true in academia, in which individual reputation is paramount (Whitley 2000). As Freeman and colleagues (2015, p. 42) note, "arguably the biggest problems collaborations must solve in order to succeed is to find ways to divide the credit." As Hagstrom (1964, p. 253) noted in the

1960s, "individuals become anonymous in large groups; talents of individual cannot be easily judged." Order of authorship, though typically indicative of the relative contributions of each author (i.e., who did more than whom), does not indicate how much more some authors did, nor how the labor was divided. Currently, evaluation and promotion and tenure committees rely heavily on letters of reference to ascertain this information. However, relying "on the word of superordinates and others on the research team reduces the independence of the individual" (Hagstrom 1964, p. 253). A more complex credit allocation system is necessary (Bikard et al. 2015, p. 6), one that goes beyond the general principles outlined in ethics codes and provides details in an explicit and public way. This is where journal policy can really help shape practice.

An important and broad implication of increasing collaboration is that inequality is intensifying, at least at the individual and institutional levels. In their study of 4.2 million articles published between 1975 and 2005, Jones et al. (2008, p. 1261) find that scientists are selecting coauthors across universities, but not across university prestige levels. Scientists at elite universities are more likely to collaborate with scientists at other elite universities, and the same homophilous tendency is seen among scientists at nonelite universities. Jones and colleagues conclude that social distance is becoming more important, just as geographic distance is becoming less important. Moreover, because multi-university publications have a higher impact relative to articles published by scholars at a single university, and researchers at elite universities are more likely to engage in such multi-university collaborations, the production of outstanding scientific knowledge is increasingly concentrated (Jones et al. 2008, p. 1261). The diffusion of scientific knowledge is also increasingly concentrated. In their study of Stanford faculty from 1997 to 2006, Rawlings and colleagues (2015) find that although more faculty are receiving knowledge from fellow faculty (i.e., adopting references from a fellow faculty member's published references), fewer are responsible for sending knowledge: The proportion of faculty sending knowledge declined from 61% in 1997 to 44% in 2006. "Star" scholars, who are more likely to be the sources of knowledge flows (Rawlings et al. 2015), also improve the quality and quantity of their collaborators' productivity (Azoulay et al. 2010). The sources of knowledge are becoming increasingly concentrated (Evans 2008) even in this interdisciplinary era.

Will these costs increase and possibly slow the trend toward team science? Probably not. Hackett (2005) and Murray (2010) show how tensions can be creative and advance the work to be done, especially if they are addressed productively and wisely. However, the identification of the costs of collaboration has certainly spurred efforts to mitigate them. Funding agencies support collaborative efforts financially. A prominent example is the NIH's Roadmap for Medical Research, launched in 2004 to transform the way biomedical research is conducted by fostering high-risk/high-reward research, enabling the development of transformative tools and techniques, filling knowledge gaps, and changing academic culture to foster collaboration. Universities are doing the same with internal funding mechanisms. Research administrators are also establishing research centers to enhance communication and reduce physical barriers to interaction (Boardman & Ponomariov 2014, Dahlander & McFarland 2013, Kabo et al. 2014), fostering mentoring and training programs within and across departments (Hackett & Rhoten 2009), and revising standards for the evaluation of interdisciplinary scholars (Boix Mansilla 2006). Moreover, the newly formed research community Science of Team Science is actively studying the elements of effective teams (Bennett & Gadlin 2012, Cooke & Hilton 2015, Stokols et al. 2008). Trust is a critical component of collaborative research (Shapin 1994, Shrum et al. 2007), as is face-to-face interaction (Hampton & Parker 2011) and a shared cognitive–emotional–interactional platform (Boix Mansilla et al. 2016), particularly for interdisciplinary collaborations.

There are several ways in which individual scientists can mitigate the costs of collaboration. They may, for example, engage in repeat collaborations by working on projects and publishing with the same set (or subset) of authors repeatedly. As Dahlander & McFarland (2013, pp. 69–70) detail, repeat collaborations have "fewer startup costs than new ones, they entail greater certainty and trust," and collaborators communicate better; this prompts "easier and more effective communication" and more "reciprocal forms of exchange." They also find that repeat collaboration (what they call persistent ties) offers greater returns on the rate of productivity and performance quality [although Inoue & Liu (2015) find that innovation can suffer, at least with respect to patents]. Individual scientists may also collaborate with like-minded others who share areas of expertise, methodological approaches, or theoretical perspectives. As elaborated in the next section, such a reinforcing style of collaboration likely improves efficiency and productivity because "returns to the time spent on tasks are usually greater to workers who concentrate on a narrower range of skills" (Becker & Murphy 1992, p. 1137). It also suggests that "sticking with one's own" is a viable and common strategy in this age of what some call hyperspecialization and fractionalization (Collins 1986, Davis 2001). When coauthors share areas of expertise, the typical costs of collaboration (e.g., coordination, control, different motives; see Shrum et al. 2007) are mitigated: Epistemological divides (Knorr Cetina 1999) are rare, shared understandings are easier to achieve (Lamont 2009), and efficient and exploitative search strategies are possible (March 1991). All of these reduce uncertainty and foster confidence in predictable and positive outcomes. And lastly, individual scientists may choose to work closer to home in terms of not only intellectual space but also geographic space, and thereby reduce or eliminate concerns about distance, lack of interaction, time zone differences, and the like—because distance still matters (Kabo et al. 2014, Olsen & Olsen 2000).

However, such reliable "succession" (Bourdieu 1975, p. 30) also has drawbacks. Namely, it does not foster broad exploration, the pooling of diverse ideas, and the development of innovative, creative ideas (Hargadon 2002, Weitzman 1998). It rarely pushes the boundaries of science in new and fruitful ways, and it may not be conducive to scientific breakthroughs and transformative science. The highest-impact collaborations are those that reach across domains, like disciplines, institutions, and countries (Bikard et al. 2015, Larivière et al. 2015a, Leahey et al. 2015, Uzzi et al. 2013), and when scientists work to mitigate the costs of collaboration, they may also be sacrificing or at least tempering these potential gains. Although collaborations between complementary specialists are rare in the field of sociology, representing only 11% of articles (Leahey & Reikowsky 2008), these boundary-spanning collaborations are especially fruitful (Leahey & Moody 2014). Again we see trade-offs.

EXTRACT 7.3

Strober, M.H. (2011) 'Difficult Dialogues: Talking Across Cultures', in *Interdisciplinary Conversations: Challenging Habits of Thought*, Stanford, CA: Stanford University Press, Chapter 3.

The Jefferson Seminar on Representation

Background

The representation seminar was the second seminar at Jefferson. The funding proposal for it, as well as the one for the first seminar, which was on consilience, was written by Jefferson's dean of arts and sciences, Joyce, a literary scholar, and its provost, Ed, a chemist. The leader of the representation seminar was Sam, a humanist who had been a participant in the first seminar and had just become an emeritus faculty member. It was Sam who chose representation as the seminar theme and explained that he wanted the seminar to explore how "reality" is represented in each discipline.

The seminar had fourteen tenured faculty: three scientists (professors of chemistry, computer science, and mathematics), four social scientists (professors of anthropology, economics, history, and sociology), five humanists (three professors from English and one each from film studies and literature), and two from the arts (a professor of music and an associate professor of fine arts). In addition, there was an artist-in-residence from the drama department. Five of the participants (slightly more than one-third) had been in the seminar the previous year. Of these, four were administrators or former administrators and one was the seminar leader. Half of the seminar faculty were women (see Appendix Table A-1).

The seminar also had nine postdoctoral fellows, more than any other seminar. Three were in the sciences (chemistry, computer science, and mathematics), two in the social sciences (sociology/anthropology and history), two in the humanities (English and American studies), and two in the arts (one in theater and one in fine arts).

Key Players

I interviewed eight members of the representation seminar (see Appendix Table A-2). The interviewees who play a part in the stories I am about to relate are mathematician Barry, studio artist Evelyn, dramatist Jane, and the seminar leader, Sam.

Mathematician Barry said he joined the seminar to learn new material and meet new colleagues. He was not particularly interested in the seminar's topic, how various disciplines represent knowledge. Artist Evelyn and dramatist Jane both said they joined the seminar because they felt isolated in their respective departments and thought the seminar could mitigate the sense of separation they had from the university as a whole. Neither was especially interested in the topic of interdisciplinarity. Sam, a professor of humanities, had had many years of experience with interdisciplinarity, including spending several years leading a multiuniversity interdisciplinary seminar.

Mathematics—The Importance of Language

Barry's participation in the representation seminar gave rise to two problems. First, seminar participants were unable to understand his presentation. Second, although he came to the seminar regularly, he never made comments or asked questions; the other participants said they were perplexed by his silence. Several people commented on the difficulty of understanding Barry's talk.

They did not have the background in mathematics to appreciate what Barry was saying, and Barry seemed to lack the skills (or motivation) to translate for them. Dramatist Jane responded to my question about whether she had experienced any difficult moments in the seminar by citing her inability to understand Barry.

> A mathematics professor who couldn't communicate ... in English. He could communicate in numbers and formulas, and it took me several, well, I think it was several hours before I understood that what he was [doing], and I was dutifully writing down everything he wrote so I could try to understand it. And when we asked questions, he would think for a long time and then say he couldn't answer it. ... That was a ... difficult period two different languages ... but ... no means of communicating.

Not only did seminar participants have trouble understanding Barry; he had trouble understanding them. In our interview, he corroborated his lack of participation.

> I was much less an active participant than I had hoped, [than I] would have liked to have been.
> Q: Any particular reason?
> A: [Pause] I think it wasn't really my medium, my natural habitat, to be in a situation like that. I hadn't gotten used to that kind of interaction. It is quite different from the way mathematicians interact.

When I asked Barry how mathematicians interact, his answer made it clear that he behaved in the interdisciplinary seminar in precisely the way that mathematicians behave in mathematics seminars, where only the presenter talks and everyone else listens, with the utmost attention, hoping to be able to follow the argument. The only interruptions are for clarification. Only after the seminar is over, perhaps weeks or months later, when participants have had time to fully digest what they heard, might they engage in private conversation with the presenter. Barry felt he could not participate in the representation seminar dialogue because he had not had time to absorb what he had heard.

Despite the fact that Barry had a clear understanding of why he didn't speak in the seminar (except when he presented), nobody else in the seminar did, as evidenced by their bewilderment about this when I interviewed them. Since no one had ever asked him the reasons for his nonparticipation, no one could help him to bridge the enormous gap between his style of thinking, presenting, and questioning and the dominant style in the seminar.

Still, although he did not participate in the discussions, Barry felt he had gotten a great deal out of the seminar. I asked him if he had it to do all over again whether he would participate. He said he would because it was so interesting and because he'd learned so much.

> Before, I really had very little sense of what, for example, a sociologist really works on and how they go about their work If I did it again, now I have a sense of that so, I might appreciate it more and participate more, be a more active participant.

One could interpret Barry's answer to mean that if he attended another such seminar, he would have a better understanding of its culture; he would understand that one cannot think as deeply and as extendedly in a humanities seminar as one does in a mathematics seminar. Whether he could actually change his behavior is a separate question.

Barry's experience shows clearly that language issues that inhibit interdisciplinary conversations go much deeper than matters of vocabulary. Barry understood the readings and he understood his colleagues' comments on them. Indeed, he found many of their comments intriguing and spent the seminar time quietly pondering them. It was the sociolinguistic system that was the problem for Barry.

Dell Hymes argues that language is embedded in a sociolinguistic system that varies across speech communities, each of which has its own rules of speaking. Unfortunately for ease of interdisciplinary conversation, the linguistic system of a discipline is taken for granted by its practitioners, who fail even to recognize that they operate in such a system, let alone understand its characteristics or realize that those characteristics differ significantly from those of their colleagues' speech communities. As a result, when faculty join an interdisciplinary group, they expect that the conversation will proceed according to the unrecognized and unacknowledged rules by which they play when talking to colleagues in the same discipline.

Hymes's careful analysis of the cultural norms of speaking in groups frames a series of questions that help to draw attention to some of the sociolinguistic difficulties in interdisciplinary settings: What are the norms of interaction? Is interrupting okay? How are turns for speaking arranged? What is said outright and what is said subtly?

I would add several additional questions to Hymes's list: How intensely may people be questioned? To what degree must participants assume a "friendly" (versus "hostile") demeanor and language toward one another and toward the presenter? Are people expected to make presentations using PowerPoint or slides, or is this seen as "too formal"? Are people expected to have video clips in their PowerPoint presentations, or is that seen as "excessive"? Is it expected that presenters will show the actual statistical results used to reach their conclusions or is a verbal summary of main findings viewed as sufficient? Are people expected to write out their presentations and read them verbatim or to speak extemporaneously from a few notes? There are also questions concerning evidence and truth claims: What constitutes a good argument? What evidence is required to reject or confirm a hypothesis?

In general, in the six seminars I investigated, the sociolinguistic system in place was chosen (without consciousness) by the leader of the seminar simply because he or she was familiar with it. There was never discussion about the ground rules of the sociolinguistic system. And in my interviews, it was only those who came from speech communities with vastly different rules who pointed out the difficulties they had adapting to the seminar style.

Commentary

Nathalie Dupin

Individuals bring different cognitive and technical skills to collaborations through communication practices (Extract 7.1), the way they relate to each other (Extract 7.2), and habits of mind (Extract 7.3). Such variety is widely associated with the complexity of teamwork where different disciplinary cultures meet, increasing the potential for tensions and conflicts among stakeholders. Yet, diversity of team members is key to tackling complex societal and environmental problems from different perspectives (Bammer, 2022). In my experience, new researchers may underestimate the skills involved in building multiple relationships in inter- and transdisciplinary collaborations or teamwork. The key readings in this chapter offer valuable insights into the competencies necessary for inter- and transdisciplinary research and the consequences of staying entrenched in a disciplinary mode of thinking. Among the findings presented in the extracts, I have chosen to confine my reflections to interpersonal dynamics through three concepts that I feel would most benefit newer researchers or those new to inter- and transdisciplinary research: learning, ethics and reflexivity.

In a discussion devoted to inter- and transdisciplinarity, there is a danger of blindness to the disciplinary backgrounds and interests of the various authors involved, thereby missing out on the richness of the perspectives offered. Broadly, Julie Thompson Klein was a scholar of interdisciplinary research and integration issues, Erin Leahey is a sociologist interested in scientific careers, and Myra Strober is a labour economist and a feminist. For completeness, I am a former research administrator (see also the commentary in Chapter 8, this book) and now PhD candidate in science and technology studies, currently researching interdisciplinary higher education. As alluded to in the chapter overview, and implicit in the range of disciplines involved, each work is underpinned by various methods, concepts and perspectives. Yet all the works here are concerned with inter- and transdisciplinary research and, more specifically, with the interpersonal dynamics at work in collaborative or team research. As a person experienced in working in teams, mostly harmoniously, I am sometimes surprised at how quickly clashes between members can happen. Sometimes there is no clash between team members but segregation, as I experienced during an exercise with a team of life sciences doctoral students, who cast me out as the 'social scientist'.

Interpersonal relationships are at the heart of efficient inter- and transdisciplinary research for developing shared understanding and goals, and facilitating interactions (Extract 7.1). As a researcher, you may already have developed a vast network of colleagues, or you may be starting networking. However, team research is a lot more unpredictable than day-to-day interactions with colleagues; it has to be sustained over time, and a lot more is at stake. Team research either requires interacting with new members or towards new research problems, maybe in different circumstances, and is, therefore, likely to involve new relational dynamics. Research relationships are complex and poorly understood, even in the case of the student–supervisor dyad (Kaasila and Lutovac, 2015). Inter- and transdisciplinary research is likely to involve multiple relationships, adding to the complexity of developing collaborative or team research with good communication, negotiation and integration skills. Without such coherence, the team is in danger of producing piecemeal research, and the project will fall short of delivering new inter- or transdisciplinary knowledge. I recognise in the work of Kaasila and Lutovac (2015) my own experience of collaborations, and I especially identify with their notion of 'expedition into shared insight', or what Klein calls 'making sense together' in Extract 7.1. Inter- and transdisciplinary research involves relationships and interactions that need to be managed so that participants can learn from each other and enable the group to overcome the challenges of team research. In collaborative work, there is much to learn about: how to communicate with other disciplines and non-academic participants, how to participate efficiently, the role of each researcher, and more. As Leahey points out in Extract 7.2, there is much to benefit from collaborations, but building effective interpersonal relationships is undoubtedly a cost – both to the collaboration in terms of time and resources and to individual members who really have to invest themselves personally.

Leahey also identifies free-riding and exploitative behaviours as potential costs to collaboration, which brings me to the second concept I wanted to reflect on in this commentary, namely ethics. More particularly, and borrowing from Etherington (2007), I believe in the importance of 'ethical research within reflexive relationships' in inter- and transdisciplinary research, not only towards potential participants or to secure the ethics committee's approval, but also between the researchers. Precisely because inter- and transdisciplinary research calls for a plurality of perspectives, an ethical stance between group members is critical to the functioning of the team. I have heard on many occasions individuals describing inter- and transdisciplinary research as 'speaking a different language', which is reflected clearly in Strober's disciplinary habits of mind and the idea of being lost in translation (Extract 7.3). This explains why so much of the literature deals with cross-disciplinary communication to enable research across academic cultures.

But conflicts can also arise in collaborative research when participants abuse their academic rank or status or privilege a hierarchical view of disciplines (Extract 7.1). While disagreements between team members can add to creativity and diversity (Bammer, 2022), actual conflict can potentially affect the whole team and jeopardise the project's outcome. When such power games are at play, communication, integration and relations suffer because of a lack of trust, as hinted by Klein and depicted in Strober's seminar (Extract 7.3). I firmly believe that remaining ethical in our relationships is a critical factor in building trust within a team. For example, giving each member – whether academic or not – the opportunity to ask questions or participate in the research process contributes to the development of a 'safe space' within the collaboration and helps build trust to invite individuals' best efforts. At a higher level, the idea of ethical relationships is equally salient to the issue of cooperation between the social sciences and humanities and science and engineering (Felt, 2014; Vienni-Baptista et al, 2022; see also the 'Introductory Essay' in this book), where 'new knowledge relations' may need to be developed to deliver innovative research (Felt, 2014).

My last reflection prompted by the readings in this chapter is the need for reflexivity in inter- and transdisciplinary research. So far, I have discussed how individual group members need to learn from each other for effective collaborative work and the importance of an ethical stance in relationships to maintain safe spaces in which members can freely contribute their best efforts. I would argue that learning and being ethical are essential aspects of knowledge production processes and are continually evolving practices that require us to be aware of, and monitor, our own actions in relation to others. Reflexivity is a widely used concept in qualitative research (Pillow, 2003) and many other disciplines (Cunliffe, 2003), yet it remains a difficult concept to grasp. Here, I refer to the act of questioning assumptions, practices and actions (Cunliffe and Jun, 2005) at the collective level to facilitate learning (see, for example, Nicolini et al, 2004). While self-awareness is important, teams need to incorporate reflexivity as a group to assess the team's progress, ensure a variety of perspectives and guard against biases such as 'habits of mind' (see Extract 7.3) that may take over the whole research process. In the words of Bourdieu (2004: 114): 'reflexivity takes on its full efficacy only when it is embodied in collectives which have so much incorporated it that they practise it as a reflex.' In agreement with Klein in Extract 7.1, the management aspect of teams and collaborations is key to ensuring that such visions become understood, shared and maintained.

In this discussion, I have focused and reflected on the interpersonal meta-skills that I see as crucial for inter- and transdisciplinary research, and as prompted by my own experiences and the three extracts from Klein, Leahey and Strober. Research support and scholar development are generally available at universities that may also offer courses such as 'handling difficult conversations', 'leadership skills' or 'values-led approach'. Relationships and interactions are something that individuals engage in, in their day-to-day life, sometimes without much reflection, and are often neglected as foundational practices in research, especially when researching with others, whether across disciplines, sectors or countries. These practices are often taken for granted and may remain unexamined but they are also challenging to apprehend because, as individuals, we are not equipped to judge them on our own. Interpersonal dynamics happen in the moment of social interaction, such as within an inter- or transdisciplinary collaboration; they are unpredictable because they are disturbed by tensions, habits of mind and personal clashes. To be effective, these relationships take time to evolve.

To conclude this discussion, my takeaway message is that inter- and transdisciplinary research are ways of knowing that rely on traditional research skills and specific competencies for communication and integration. In addition, collaborations and teamwork rest on efficient interpersonal interactions that involve:

- *Mutual learning:* Teams learn much from their participants and from their research, but members also learn by arriving at shared understandings, language and goals. They learn through each other that many perspectives and solutions may exist and are valid, and learn to keep an open mind.
- *Responsibilities:* Remaining ethical in relationships helps to build safe spaces and promote trust within the team, enabling strong relationships and learning. Equally, each team member must participate in the research process to avoid freeriding.
- *Meta-skills:* Collective reflexivity can be learned and can become a routine practice that allows guarding against biases, monitoring team evolution and communications, as well as the consideration of issues of equality, diversity and inclusion – that reinforce the ethical aspect of the relationships. A host of meta-skills, such as being creative, dealing with uncertainty and remaining open-minded, are also a requisite for inter- and transdisciplinary research, and they can all benefit from a level of reflexivity.

Finally, it is worth repeating that there is no one right way to be interdisciplinary (Fletcher and Lyall, 2021: 54–5; Vienni-Baptista et al, 2022). Inter- and transdisciplinary research are a subset of academic practices and modes of knowledge production. Just as there are many ways of doing ethnographic research, for example, inter- and transdisciplinarity imply multiple practices that may mean different things to different individuals or institutions. The reflections I have presented here have led me to discuss relationships and interactions and how they can benefit from mutual learning, an ethical stance and reflexivity. I have found some of these competencies lacking in past collaborations, and I believe researchers can enhance their abilities to participate in inter- and transdisciplinary research by paying attention to these aspects of research that are often overlooked, both in practice and in existing scholarship.

References and further reading

Bammer, G. (2022) 'Understanding diversity primer: 1. Why diversity?', Integration and Implementation Insights, Blog, 21 April, https://i2insights.org/2022/04/21/why-diversity

Bourdieu, P. (2004) *Science of Science and Reflexivity*, Cambridge: Polity.

Cunliffe, A.L. (2003) 'Reflexive inquiry in organizational research: Questions and possibilities', *Human Relations,* 56(8): 983–1003, https://doi.org/10.1177/00187267030568004

Cunliffe, A.L. and Jun, J.S. (2005) 'The need for reflexivity in public administration', *Administration & Society*, 37(2): 225–42, https://doi.org/10.1177/0095399704273209

Etherington, K. (2007) 'Ethical research in reflexive relationships', *Qualitative Inquiry*, 13(5): 599–616, https://doi.org/10.1177/1077800407301175

Felt, U. (2014) 'Within, across and beyond: Reconsidering the role of social sciences and humanities in Europe', *Science as Culture*, 23(3): 384–96, https://doi.org/10.1080/09505431.2014.926146

Fletcher, I. and Lyall, C. (2021) 'Stem Cells and Serendipity: Unburdening Social Scientists' Feelings of Failure', in D. Fam and M. O'Rourke (eds) *Interdisciplinary and Transdisciplinary Failures: Lessons Learned from Cautionary Tales*, Abingdon: Routledge, Chapter 3.

Kaasila, R. and Lutovac, S. (2015) 'Developing research relationships toward a learning partnership', *Scandinavian Journal of Educational Research*, 59(2): 177–94, https://doi.org/10.1080/00313831.2014.904415

Nicolini, D., Sher, M., Childerstone, S. and Gorli, M. (2004) 'In Search of the "Structure that Reflects": Promoting Organizational Reflection Practices in a UK Health Authority', in M. Reynolds, R. Vince, J.A. Raelin, M.A. Welsh and G.E. Dehler (eds) *Organizing Reflection*, Abingdon: Routledge, Chapter 6.

Pillow, W. (2003) 'Confession, catharsis, or cure? Rethinking the uses of reflexivity as methodological power in qualitative research', *International Journal of Qualitative Studies in Education*, 16(2): 75–196, https://doi.org/10.1080/0951839032000060635

Vienni-Baptista, B., Fletcher, I., Lyall, C. and Pohl, C. (2022) 'Embracing heterogeneity: Why plural understandings strengthen interdisciplinarity and transdisciplinarity', *Science and Public Policy*, 49(6): 865–77, https://doi.org/10.1093/scipol/scac034

Supporting Collaborative Researchers

Chapter overview

Universities have traditionally been organised around discipline-based structures and norms. Fostering an environment that encourages and supports collaboration across disciplines may require cultural, procedural and structural changes. Developing such support requires better understanding of the specific expertise and skills required to undertake collaborative research, and how researchers can acquire such skills.

Expertise in research integration and implementation is an essential component of tackling complex societal and environmental problems. Much of the expertise is tacit and is used intuitively by researchers, practitioners and funders. Gabriele Bammer and her co-authors come from diverse scientific communities. In this article (see Extract 8.1), they explore three questions: (1) When is expertise in research integration and implementation required?; (2) Where can expertise in research integration and implementation currently be found?; and (3) What is required to strengthen expertise in research integration and implementation? We have extracted sections that consider the second and third of these questions. The authors conclude that building a knowledge bank and a coalition of researchers and institutions will ensure that this expertise and its application are valued and sustained.

Research to successfully address complex and multidimensional problems demands effective leadership for inter- and transdisciplinary institutions. Based on the diverse experiences of 20 institutional leaders, the article by Christopher Boone et al (Extract 8.2) shows ways to cultivate appropriate leadership qualities and skills, especially the ability to create and foster vision beyond the status quo, collaborative leadership and partnerships, and shared culture, among others. A valuable framework that articulates the major facets of leadership in inter- and transdisciplinary organisations (learning, supporting, sharing and training) constitutes the main contribution of this article, and Extract 8.2 addresses leadership qualities and skills.

In their chapter for *The Oxford Handbook of Interdisciplinarity*, Stephanie Pfirman and Paula Martin explore approaches to interdisciplinary scholarship in comparison to disciplinary traditions in higher education (Extract 8.3). They investigate the particular challenges of interdisciplinary research, teaching and service for scholars throughout a scholarly career. Ideas to support interdisciplinary faculty are presented, from the creation of the position to the point of hire, and through a career timeline to tenure and post-tenure review. Special challenges in interdisciplinary scholarly productivity, scholarly recognition, evaluation, promotion and funding are examined, and we have selected sections in Extract 8.3 outlining the institutional support needed to address these challenges.

EXTRACT 8.1

Bammer, G., O'Rourke, M., O'Connell, D., Neuhauser, L., Midgley, G., Klein, J.T., Grigg, N.J., Gadlin, H., Elsum, I.R., Bursztyn, M., Fulton, E.A., Pohl, C., Smithson, M., Vilsmaier, U., Bergmann, M., Jaeger, J., Merkx, F., Vienni-Baptista, B., Burgman, M.A., Walker, D.H., Young, J., Bradbury, H., Crawford, L., Haryanto, B., Pachanee, C.a., Polk, M. and Richardson, G.P. (2020) 'Expertise in research integration and implementation for tackling complex problems: When is it needed, where can it be found and how can it be strengthened?', *Palgrave Communications*, 6(1): 5.

Question 2: where can expertise in research integration and implementation currently be found?

We have identified three major realms where expertise in research integration and implementation can be found and how they correspond (or not) to communities of researchers. First, some researchers apply specific approaches to tackling complex societal and environmental problems, such as interdisciplinarity, systems thinking, and action research. These approaches have coalesced around particular ways of understanding and operationalising research integration and implementation. Each community practising a specific approach is largely independent of the others.

Second, some researchers develop case-based experience without reference to specific approaches and, by moving from one problem to another, progressively build useful know-that and know-how expertise (often tacit) in research integration and implementation, along with interactional expertise. From time to time they may incorporate know-that and know-how developed by others into their practice. Unlike researchers using specific approaches, researchers drawing primarily on case-based experience tend not to be organised into communities around expertise in research integration and implementation, although they may be organised into communities around the problems of interest.

We recognise that the communities using specific approaches to complex societal and environmental problems also work on cases (see for example Fulton et al. (2014) for a case using complex systems science and Neuhauser (2018) for a case using transdisciplinarity), but that is tangential to the point we make here.

The final source of expertise comes from researchers who investigate an element of research integration and implementation and who are not aligned with either of the other realms. We focus here on two examples—researchers interested in unknowns and those interested in innovation. In both cases, researchers come from various disciplinary and professional backgrounds. These examples differ in the strength of the associated community (measured by regular conferences and publishing in specific journals), which is weak in the case of unknowns and stronger for innovation. In both cases, interest in unknowns or innovation is not specifically focused on complex societal and environmental problems and the relevance of their insights to research integration and implementation may not be immediately obvious.

Of course, the three realms do not have hard boundaries and researchers may identify with different realms at different times in their careers. The point of identifying these three realms where expertise in research integration and implementation exists is to highlight both the existing fragmentation, as well as which veins need to be tapped into to draw together what is already available, especially in relevant know-that and know-how contributory expertise, and to illustrate where interactional expertise and tacit expertise are important.

Specific approaches. One rich source of insights into expertise in research integration and implementation can be found in what we call specific approaches. The lists in Box 1 come from a sub-group of the authors who drew on several centuries of combined experience and scholarship, as well as their roles in helping develop some of these approaches (marked with an 'a'). The first column in Box 1 records 14 approaches that provide a wide range of expertise across both research integration and implementation. These approaches include action research, integrated assessment and post-normal science. The ten approaches in the second column provide a subset of expertise in research integration and implementation. Some provide expertise in research implementation only (change management, impact assessment, impact evaluation, implementation science, K* and policy science). Others provide expertise across both research integration and implementation, but only for a specific set of activities, rather than the broad range of expertise provided by the approaches listed in the first column. In particular, three approaches provide expertise in decision making and/or dealing with risk (decision making under deep uncertainty, decision sciences and risk analysis) and one in collaboration (science of team science). Multiple related approaches are listed under one specific approach in each column, namely systems thinking and K*. Many other approaches encompass various schools of thinking as well, but these have not split into separate but related approaches. While we have listed all the specific approaches we are aware of, we anticipate that this list is not complete and that there are elements which could be contested. Further, we have not aimed to be comprehensive in the cited references, but rather have provided a major work (occasionally more) as a starting point for those interested in learning about each approach.

Box 1 | Specific approaches to tackling complex real-world problems, divided into those that provide wide-ranging expertise in both research integration and implementation and those that provide expertise in a subset of research integration and implementation activities

Approaches providing wide-ranging expertise across both research integration and implementation

---

Action research[a,1]
Complex project management[a,2]
Complex systems science[3]
Design science[a,4]
Integrated assessment[a,5]
Integration and implementation sciences[a,6]
Interdisciplinarity[a,7]
Mode 2[8]
Operational research[9], including community operational research[a,10]
Post-normal science[11]
Sustainability science[a,12]
Sustainability transitions[13]
Systems thinking[a], including systems analysis[a,14], systems engineering[15], the viable system model[16], systems failure[17], soft systems thinking[18], critical systems thinking[19], system dynamics[a,20] (including participatory system dynamics[a,21]) systemic intervention[a,22]
Transdisciplinarity[a,23]

1 Bradbury, 2015
2 Cicmil et al., 2009
3 Mitchell, 2009
4 Simon, 1996
5 Rotmans and van Asselt, 1996
6 Bammer, 2013
7 Klein, 1990; 2010
8 Gibbons et al., 1994
9 Taha, 2017
10 Johnson, 2012; Midgley and Ochoa-Arias, 2004
11 Funtowicz and Ravetz, 1993
12 Clark and Dickson, 2003
13 Loorbach et al., 2017
14 Miser and Quade, 1985; 1988
15 Hall, 1962
16 Beer, 1984
17 Fortune and Peters, 1995
18 Ackoff, 1981; Checkland, 1981; Churchman, 1979
19 Flood and Jackson, 1991; Flood and Romm, 1996
20 Forrester, 1961; Sterman, 2000
21 Richardson and Andersen, 2010; Vennix, 1996
22 Midgley, 2000
23 Bergmann et al., 2012; Hirsch Hadorn et al., 2008; Jahn et al., 2012; Vilsmaier et al., 2017

---

[a] Approaches that members of the authorship group have helped develop

---

Approaches providing expertise for a subset of research integration and implementation activities

Research implementation only
Change management[24]
Impact assessment[25]
Impact evaluation[26]
Implementation science[27]
K* (KStar) including Knowledge brokering, Knowledge exchange, Knowledge management, Knowledge mobilisation, Knowledge transfer, Knowledge translation[28]
Policy science[29]

Specific subset of research integration and implementation
Decision making under deep uncertainty[30] Decision sciences[31]
Risk analysis[32]
Science of team science[a,33]

24 Nauheimer, 1997
25 Therival and Wood, 2018
26 Gertler et al., 2016
27 Eccles et al., 2009
28 Shaxson et al., 2012
29 Cairney and Weible, 2017
30 Marchau et al., 2019
31 Kleindorfer et al., 1993
32 Aven, 2012
33 National Research Council, 2015

[...]

Question 3: what is required to strengthen expertise in research integration and implementation?

Key to strengthening expertise in research integration and implementation is to make it readily identifiable and accessible.

As little effort has gone into documenting such expertise, it is largely invisible and unrecognised. Further, as we have shown in the previous section, it is also highly fragmented. It is, therefore, currently much easier for researchers and teams to 'reinvent the wheel' by duplicating know-that and know-how than to find build on and improve existing expertise.

Identifying expertise and overcoming fragmentation are therefore critical, requiring both an inventory and an organisational framework that promotes accessibility. Both requirements could be achieved by building a dynamic, shared knowledge bank and here we outline what is involved. The scale of a knowledge bank would be considerably greater than a toolkit, and indeed, many toolkits would be included in the knowledge bank. In order to avoid consignment to a graveyard of integrative databases, atlases, and knowledge compendiums that have not gained traction, the knowledge bank requires wide-ranging support from the realms where expertise currently resides. Building a knowledge bank therefore also involves building a coalition of key communities and teams. Such a coalition will, through the process of building the knowledge bank, provide an authoritative voice about expertise in research integration and implementation, ensuring that it is properly valued. In turn this will improve assessments of integration and implementation in research tackling complex societal and environmental problems, including in tenure and promotion applications, funding proposals, outcomes of research projects, and outputs of inter- and transdisciplinary centres and other institutions. We review three key challenges involved in building a knowledge bank: compiling existing expertise, indexing and organising the expertise to make it widely accessible, and understanding and overcoming the core reasons for the existing fragmentation. For each we also highlight some current-positive trends. The aim is to provide an indication of the effort that is required, especially from those in the three realms who need to be involved for a knowledge bank to be a long-term success.

Compiling existing expertise. Compiling existing expertise is a major task and we have aimed to lay foundations in this article by addressing three key issues:

- defining expertise
- identifying tasks requiring expertise in research integration and implementation
- describing expertise, which also needs to incorporate illustrative examples and guidance for when the expertise is appropriate (relating back to the tasks requiring expertise).

In focusing on the third of these—describing expertise—we identify five major challenges. One immediate challenge is finding relevant expertise and illustrative case studies, many of which are not documented in either the published or grey literatures. Those that are documented are widely dispersed and not easy to locate.

Second, developing guidance to link expertise to tasks is hampered by lack of evidence about key aspects of most know-that and know-how, including strengths and weaknesses, effectiveness, or how well the expertise can be adapted to differing circumstances.

Third, elements of expertise in research integration and implementation will be characterised differently within and across the three realms. This is affirmed by our experience of working together as authors, where we found, for example, different framings of power relations and how they should be addressed. As well as identifying differences, it is essential to recognise when particular know-that or know-how expertise, as used by different groups, is essentially the same.

Fourth, those who currently have expertise in research integration and implementation come from very different backgrounds, so that expertise that is self-evident to some will be a revelation to others. For example, the importance of understanding and managing power relations will be obvious to those originally trained in the social sciences, but will be a topic some others rarely consider. Similarly, the possibility of non-linear relations between two causal factors will be self-evident to those originally trained in the physical sciences, but new to many others.

As hinted at in these examples, there are many areas where expertise in research integration and implementation intersects with expertise in existing disciplines. How this is dealt with in deciding what expertise to include in a compilation of know-that and know-how for research integration and implementation requires careful consideration.

Finally, the biggest challenge in compiling and assessing expertise, as well as writing guidance notes, is that no single individual or group has experience across the three realms, or even a significant subset. Instead, a more labour-intensive and time-consuming process is necessary, requiring individuals and groups from the different realms to work together globally to understand each other's contributions, before being able to undertake the requisite sifting, assessing and provision of guidance.

It is beyond the scope of this article to do more than flag these matters. Nevertheless, we see the challenges as both intellectually exciting and practical, although far from straight-forward to manage.

Building on positive trends. Growing recognition of connections across different specific approaches is a positive trend, especially for identifying the broad range of expertise to be included in a knowledge bank. For example, the *Oxford Handbook of Interdisciplinarity* (Frodeman, 2017) includes information on transdisciplinarity, systems thinking, design science, team science, sustainability science, integration and implementation sciences, and innovation, along with its primary focus on interdisciplinarity. Similarly, the *Sage Handbook of Action Research* (Bradbury, 2015) provides chapters on systems thinking and integration and implementation sciences. In the same vein, the journal *GAIA* has published a series of Toolkits for Transdisciplinarity (Bammer, 2017) and is now issuing a series of Frameworks for Transdisciplinary Research that are drawing connections, not only to other specific approaches such as systems thinking, change management and integration and implementation sciences (e.g., Cabrera and Cabrera, 2018 on systems thinking), but also to dialogue and collaboration methods from case-based experience (e.g., Bammer, 2016b; McDonald et al., 2009)

Indexing and organising the expertise and making it widely accessible. Compiling expertise is not enough. The knowledge bank needs to be organised in a way that makes expertise easy to find by a wide range of interested individuals, teams and communities of practice, as both contributors and users. Multiple entry points are required, providing access that is intuitive and welcoming, and that accommodates the different ways of understanding and tackling complex societal and environmental problems described in this article.

Finding an effective way to index expertise in the knowledge bank is key. This formidable challenge calls for collaboration with information scientists to build on lessons from existing knowledge banks and cyberinfrastructures, including Wikipedia (2019), Dryad Digital Repository (2019), Open Biological and Biomedical Ontology (OBO) Foundry (2019), Gene Ontology Resource (2019), as well as the Long-Term Ecological Research Network (2019) and the National Ecological Observatory Network (2019). In addition, Leonelli and Ankeny (2015, p. 703) highlight a 'tension between the stability and the flexibility of classificatory categories,' which need both consistency over time and to be adapted and updated so that they 'mirror the research practices and knowledge of their users'.

A host of other practical matters must also be resolved, including funding, maintaining long-term integrity and interoperability, establishing meta-data standards, encompassing burgeoning variations in nomenclature, determining who can contribute and how, providing credit and other incentives to contribute, and establishing standards for assessment of contributions. Again, how such challenges are addressed by existing successful repositories will be instructive. For example, Wikipedia (2019) is the best-known online reference website and provides lessons about a resource written collaboratively by volunteers. Its open authorship and editing policy mean new content can be easily created and all content can be kept up-to-date, but it also means accuracy, rigour and indexing can be inconsistent.

Building on positive trends. A positive trend to build on is the activities of Integration and Implementation Sciences (i2S; Bammer, 2013), which is testing ideas relevant to developing an ontology. The i2S frame consists of three domains: synthesis of disciplinary and stakeholder knowledge, understanding and managing diverse unknowns, and providing integrated research support for policy and practice change. It also addresses five questions: For what and for whom? Of what? How? Context? Outcome? These questions encompass specific knowledge and skills in integration and implementation, such as framing, modelling and co-production. In the aggregate, these components aim to provide a structure or ontology for indexing, codified in a repository of resources (Integration and Implementation Sciences, 2019c).

An important caveat to organising expertise into a knowledge bank is the need for congruence with the challenges of dealing with complex problems. The multiple facets and interactions of complexity, and the creativity required to deal with them, cannot be ignored or downplayed in favour of indexing requirements. Indeed, the challenges of building a knowledge bank to strengthen expertise in research integration and implementation are likely to stimulate further development of information sciences towards systems that can deal effectively with complexity.

Understanding and overcoming the core reasons for the existing fragmentation. Overcoming fragmentation of expertise in research integration and implementation requires understanding why it exists and what forces maintain it. Separate explanations offer insights into fragmentation of expertise codified by various specific approaches and of expertise developed by case-based experience that is independent of specific approaches.

Specific approaches can be seen as small 'tribes' around particular 'territories', to use concepts and terms made popular by Becher (1989) in his analysis of the construction of boundaries that differentiate disciplines. By their nature, tribes organise around certain journals and conferences, and their members become reviewers for each other's work. More broadly, members of a tribe are identified by 'traditions, customs and practices, transmitted knowledge, beliefs, morals and rules of conduct, as well as their linguistic and symbolic forms of communication and the meanings they share' (Becher, 1989, p. 24). Tribal differences among specific approaches were evident in our interactions at the 2013 conference (Integration and Implementation Sciences, 2019a) and in our deliberations as an authorship group.

As an aside, Becher's insights can be further invoked to envision that building a knowledge bank is an exercise in reviewing a territory in order to formalise permeable boundaries 'open to incoming and outgoing traffic' (1989, p. 37) with the aim of rendering accessible all elements of expertise required to tackle complex societal and environmental problems. Indeed, it was the interaction of our individual 'tribes' that exposed our group to rich 'intra-tribal' knowledge and reinforced our joint commitment to developing a knowledge bank.

The fragmentation of expertise developed by case-based experience that is independent of specific approaches can be understood through the expositions on mode 2 research underpinned by transdisciplinarity (in its generic rather than specific sense) by Gibbons et al. (1994) and on interdisciplinarity (also in its generic sense) by the US National Academies (National Academy of Sciences, National Academy of Engineering and Institute of Medicine, 2005). Both publications emphasised that, at the time they were written, relevant expertise was diffused in a loose way, typically transmitted by experienced researchers moving to new problems rather than through institutionalised reporting in professional journals. This mode of diffusion continues to be typical among those focused on case-based experience, as described earlier.

Both forces of fragmentation—multiple small tribes and loose diffusion—are reinforced by a third force combining high academic workload and publication pressure (Kinman, 2014; Kinman and Jones, 2003). This combination severely limits the time researchers have to look beyond their own tribe or accumulated experience to find useful know-that and knowhow expertise, especially when doing so requires familiarity with a wide array of literature and multiple specialised 'languages.'

Building on positive trends. At least three encouraging trends are overcoming fragmentation. First, 'borrowing' (i.e., taking concepts and methods from one discipline into another) has long been an acknowledged feature of interdisciplinary research (Klein, 1990). In addition, it occurs across specific approaches. For example, the interdisciplinary tool "Toolbox' dialogue method to uncover research worldviews' (see Box 2) has been incorporated into an online toolkit of co-production methods for transdisciplinary research (td-Net, 2019). Borrowing is also starting to link case-based experience with specific approaches. Population health research, for instance, is adapting ideas from complexity science (Long et al., 2018; Thompson et al., 2016).

A second trend is exemplified by the Integration and Implementation Insights blog (2019), which is a conduit for linking researchers, regardless of tribe or problem tackled, allowing them to share integration and implementation expertise in easy-to-read form. This is in line with the community building identified by Leonelli and Ankeny (2015) as an essential component of developing a large-scale repository.

A third trend builds on the long-standing practice of establishing dedicated centres, employing a range of disciplinary experts, to address complex societal and environmental issues. The heads of such organisations are forming an authoritative leadership group to ensure funders and research policy makers understand, value and support expertise in research integration and implementation (Network of Interdisciplinary and Transdisciplinary Research Organisations—Oceania, 2019; Palmer, 2018).

EXTRACT 8.2

Boone, C.G., Pickett, S.T.A., Bammer, G., Bawa, K., Dunne, J.A., Gordon, I.J., Hart, D., Hellmann, J., Miller, A., New, M., Ometto, J.P., Taylor K., Wendorf, G., Agrawal, A., Bertsch, P., Campbell, C., Dodd, P., Janetos, A. and Mallee, H. (2020) 'Preparing interdisciplinary leadership for a sustainable future', *Sustainability Science*, 15(6): 1723–33.

Cultivate appropriate leadership qualities and skills

Leaders of ITD organizations need the qualities that make any leader successful—creativity, humility, open-mindedness, long-term vision, and being a team player. In addition to these general qualities, ITD leaders require skills and attributes that are specific to inter- and trans-disciplinary interactions and that have the capacity to be transformative with real-world impacts. ITD leaders often must be more persuasive than other leaders to convince researchers to follow the unsettled and novel pathways of ITD research. Qualities that have been most transformative in our own journeys as leaders are the ability to create and foster: vision beyond status quo, collaborative leadership and partnerships, shared culture, communications to multiple audiences, appropriate monitoring and evaluation programs, and perseverance. It is important to note that these leadership qualities, skills, and attributes evolved over time. We did not begin our positions with each of these at hand; rather, as our roles and institutions grew, so did our leadership in these areas. Often, no individual has all of these qualities so it is also important to build a team that incorporates the full suite of these abilities.

Vision beyond status quo

Sustainability necessitates long-term vision that goes beyond the status quo (Matson et al. 2016). The complexity and scale of the challenges we confront require working and planning at time scales longer than the tenure of individual leaders. ITD leaders need the ability and creativity to see beyond existing conditions to imagine what is possible, what is needed, and how to get there, while integrating multiple stakeholder insights. We have operated in institutions that are sometimes slow to move and hesitant to change, yet we laid out strategic long-term plans that defied existing structures to facilitate the ITD goals we articulated. Ashoka Trust for Research in Ecology and the Environment (ATREE) in India provides an example of the vision and evolution required to move beyond the status quo (Box 3).

Collaborative leadership and partnerships

Leadership is a multidimensional process. It is important to know how to share leadership and to support the many roles required for sustainability work. Designated leaders must sometimes act as supporters, or as champions outside the organization. Appreciating and practicing different roles is a key cultural habit for leaders of ITD organizations. In some circumstances, ITD leaders must act as facilitators, 'de-centering' the role of academia to effectively prioritize the voices, concerns, and ideas of diverse stakeholders (Alonso-Yanez et al. 2019). Shared leadership may mirror necessities within ITD centers. Because of the multiplicity of leadership attributes, a team of more than one leader may be appropriate. The shared leadership model—as for example practiced by ZTG in TU Berlin and by the Wrigley Institute at ASU (Box 1)—also supports the idea of non-hierarchical working-structures, raising the credibility that partners outside of academia are fully accepted for their specific knowledge and perspectives.

Effective collaboration can catalyze problem analysis and address the broad range of elements that must be considered. Collaborative methods can be central for improving use of natural resources shared by society (Talley 2016) while also enhancing governance and accountability. Nevertheless, it is important to consider how and when to collaborate with partners. There is a tendency to want to partner with everyone who is interested, particularly in sustainability where the challenges are complex and sense of urgency is strong. However, in our experience, the most effective leaders have developed clear processes for assessing whether to partner and how to measure success of partnerships. There are transaction costs to engaging partners as every partnership is a decision to allocate time and money. If not done carefully, partnerships can drain resources, taking intellectual and financial capital from other more fruitful activities. Before engaging with partners, it is important to ask key questions: Are the partner's objectives and proposed activity aligned with our strategy and operational plans? Can we establish and commit to a clear governance structure and resourcing? Is there enthusiasm from faculty and researchers? Is the proposed engagement intellectually interesting and impactful in the field? When the answers are yes, strong leaders invest to build participation, trust, excitement, and outcomes. Two examples of effective partner engagement are described in Box 4.

If an ITD organization identifies a strategic partner, it is important to engage them as much as possible from the beginning of the research process (Herrero et al. 2019). However, such participatory processes have challenges that need to be crystal clear to everyone from the outset, thereby avoiding frustrations from results that might not meet expectations (Stokols 2006; Disterheft et al. 2015). Clear articulation of the possible trade-offs between the scientific ideas and participatory methods is important to establish. A transparent set of scientific tools, visualized well across research phases, and a clear integration of different ways of expressing knowledge, including the follow-up of the results and the feedback to the stakeholders or to the practitioners, are of central importance (Mielke et al. 2017). Effective stakeholder engagement requires open access to data and knowledge so that key information is not restricted to the academic team members (Kondo et al. 2019). This approach provides informed options for decision processes while also using feedback from stakeholders to advance a specific research agenda. The development of the research or solution should be co-planned with stakeholders as this facilitates a way to effectively design and to measure outcomes. Determining outcomes with stakeholders increases the chance that results will be taken seriously and be implemented, while also incentivizing communities to help with gathering data (Heinzmann et al. 2019). However, lack of a concrete framework or model for carrying out a transdisciplinary sustainability project can increase potential for failure or reduce effectiveness of implementation (Smetschka and Gaube 2020). The risk associated with failing to meet anticipated objectives can be minimized by regularly revisiting goals and progress with all interested parties within an agreed upon evaluation framework (Williams and Robinson 2020; Turner and Baker 2020).

Shared culture

Because sustainability and ITD science are relatively new, attention to culture is crucial for future leaders (Longino 1990; Johnson and Xenos 2019). Culture includes norms and habits of mind that affect problem selection, research approaches, pathways of application (Pickett et al. 2007) and adapted solutions. Norms can limit or promote specific research and outcomes. Indeed, the traditional culture of science has promoted narrow disciplinary and academic outcomes (Capra 1983). Even tacitly adopting a familiar scientific culture may thwart the interdisciplinarity that sustainability requires.

Culture usually exists in the background, yet to succeed, leaders of ITD organizations must promote a new scientific culture that values and promoted ITD research and activities. They may have to guide their organizations through articulating and establishing new norms, finding ways to reward appropriate collaborative behaviors, and discouraging lapses into cultural norms of a narrow disciplinary past (Brown et al. 2019). Among the most significant cultural features supporting ITD success is a sharing attitude. This feature may be difficult for those trained in science as an individual, rule-based pursuit. In particular, the traditional idea that an individual researcher owns data can impede robust ITD research (Willig and Walker 2016). Consequently, sharing data in clear, well-documented, understandable formats is an important cultural norm for interdisciplinarity and transdisciplinarity.

Communications with multiple audiences

Communication is respectful listening coupled with clarity of exposition. Oral, written, quantitative, and visual modes may be combined in many ways. Conducive places for discussion, scheduled and serendipitous meetings, and access to multiple tools are all parts of effective communications in ITD organizations. Effective communication requires deep respect for other ways of knowing and social practices, especially as ITD endeavors engage increasingly diverse stakeholders. Because sustainability problems are complex, successful ITD leaders find it helpful to have a clear understanding of the logic of constituent or partner institutions and the incentives that drive stakeholders and find ways to mediate, resolve conflicts, and develop common ground priorities (Barrett et al. 2019).

Effective communication within the organization is also required to build and maintain networks uniting disciplinary expertise for ITD challenges. Communication with senior leadership of larger organizations that may host ITD centers is required to sustain buy-in while minimizing institutional friction. Leaders should adopt a variety of participation methods to integrate local expertise. Communication requires the ability to convene and engage across disciplines, to convince others, and understand how to excite researchers to participate in ITD when doing so is outside their norms (Box 5).

Appropriate monitoring and evaluation

Properly evaluating ITD research remains a challenge. It may be tempting to set over-ambitious goals. Failure to achieve such goals demotivates researchers, distances stakeholders, and disappoints funders and clients of ITD organizations. Some examples of overpromising include fundraising across too broad a scope of activities, with none funded adequately; trying to do too many things, which leads to 'dropped balls' and disappointed partners; priming junior faculty for leadership, when such positions are not available; and relying on students to produce deliverables, but not informing the funder that this necessarily includes an education component that differs from a consultancy. Back-up support also needs to be available if students fail to complete a project. Ambitious goals can be valuable in motivating innovative ITD work, but appropriate expectations need to be set from the beginning and revisited frequently with internal and external stakeholders. Establishing a flexible, dynamic evaluation and monitoring framework as close as possible to the beginning phases of programs can greatly assist the management of ITD programs, freeing up time for leaders to pursue other responsibilities. In addition to evaluating program outputs and outcomes, the framework should evaluate the effectiveness of ITD processes themselves so that learning and development can take place in ITD teams (Holzer et al. 2018).

Perseverance

As sustainability programs and ITD research inherently challenge the status quo, effective leaders must be able to articulate a shared strategy and persevere against a tendency to regress to traditional, disciplinary approaches. The normative, practical nature of sustainability, its breadth of concerns, and its shifting or inexact definitions can invite skepticism from established scientific disciplines. The tendency for scientists to believe their own disciplines have higher value than other disciplines can also fracture ITD programs. All of these dynamics are acute in the early days of ITD program development.

Leaders who persevere and continuously communicate the value and role of ITD programs and research provide time for skepticism to erode, for disciplinary scientists to develop empathy for other ways of knowing, and for the creation of shared research, education, and outreach products that demonstrate the value of ID and TD (Kelly et al. 2019). Examples from Columbia University's Earth Institute, Arizona State University, and the University of Minnesota's Institute on the Environment illustrate the necessary perseverance around the establishment of new structures and celebration of their achievements, whereas the example from Baltimore Ecosystem Study (BES) illustrates perseverance within team processes (Box 5).

Resources for success

Resources needed to enable success in positions of leadership within ITD organizations fall into five categories: (1) intellectual resources; (2) institutional policies; (3) financial resources; (4) physical infrastructure; and (5) governing boards. First, leaders need to build and sustain mechanisms for recognizing and engaging intellectual expertise outside the disciplinary academic discourse (Bammer et al. 2020). This includes engaging all partners—those within one's home institution, other academics, and a broad array of stakeholders. Such engagement elicits new ideas, perspectives, and initiatives, contributing to the dynamism that is so important to ITD research. Tapping outside experts for short engagements through visiting appointments, internships, fellowships, post-docs, speakers, or program evaluators provides concentrated value and broadens reach and scope without the long-term budget commitments of adding permanent staff (Trimble and Plummer 2019).

Secure funding to support early career researchers, including doctoral students, post-doctoral fellows, and junior faculty is central for the longevity and success of ITD research. Many junior scholars, some trained in ITD, are attracted to the mission-oriented nature of ITD programs and institutes. They want to help solve sustainability problems and need roadmaps to consult. Traditional departmental training will not be sufficient to succeed in ITD scholarship without strong mentoring, explicit incentives to engage, and guidance on best practices. Graduate students and post-doctoral fellows should be given opportunities to share leadership, especially when their ITD training can facilitate multi-investigator and stakeholder projects that involve individuals with traditional, disciplinary training or single-issue agendas (Fam et al. 2020).

Second, leaders must be aware of the role of institutional infrastructure and how to foster policies that result in collaborative relationships, non-traditional outputs and outcomes, engagement with practitioners, celebration of ITD work, and career progression from recruitment to promotion. Columbia University's Earth Institute, for example, developed practice-oriented guidelines for appointment and promotion for its research scientists, with explicit guidance on new metrics and criteria for activities outside the scope of traditional research and how to judge them. Spokespersons for ITD must not be seen as competing for funds within the organization but as adding value to existing programs. Linking ITD activities to the core culture of the institution can promote ITD work. As an example, courses co-taught by faculty from different disciplines or courses co-taught by tenured faculty and industry or non-profit professionals can lead to the co- production of novel approaches to solving topical, real-world problems.

Third, leaders need to operate based on the reality that many ITD research organizations are soft-money institutions. Long-term grants for ITD research are rare, so developing nimble ways to leverage limited budgets is critical. Experimenting with different seed funding for interaction and collaboration, such as those tied to specific outputs, can help expand into larger programs and broaden participation. Buying out faculty time or borrowing individuals for part of a year for leadership or collaborative activities can relieve constrained funding. Utilizing non-financial resources, such as staff time for proposal support, project management, or communications assistance, can also attract ITD participants from across and between institutions (Cundill et al. 2019).

However, it is important to be aware that proponents of disciplines may be openly hostile to ITD programs because they see them as direct competitors for funding. Attempts to compensate by 'buying' contributions from researchers in discipline-based departments are not always successful. Short-term income generation and time pressure are often achieved at the expense of longer term relationship building. Some organizations have found endowments to be key in allowing them to function, but maintaining a funding stream through endowments can bring its own challenges, depending on investment returns and broader economic conditions.

Fourth, the physical place and space of an ITD organization is vitally important. Co-location of scholars from different disciplines sparks serendipity—encouraging the hallway conversations and spontaneous brainstorming over coffee breaks—that inspires ITD work and reduces the need for formal meetings, seminars, and workshops (Lyall 2019). Where co-location is not possible, technology to engage distant partners electronically is an important aspect of the physical place. Co-location with external stakeholders can generate easy access to policymakers and facilitate the co-production of knowledge and solutions to real sustainability problems. One example is the Sustainable Cities Network, housed in the ASU Wrigley Institute, which brings together sustainability officers and other practitioners from municipalities and tribal governments from across the State of Arizona (https://sustainability.asu.edu/sustainable-cities/). The network identifies real-world sustainability problems as opportunities for research, education, and outreach. An example of an established ongoing program that resulted from this network is Project Cities, which links courses from across Arizona State University to solve specific community solutions, with monetary and other support from the participating cities (https://sustainability.asu.edu/project-cities/).

Finally, trustees, governing boards, or members of advisory bodies are important ITD resources. Supportive boards can advocate across their networks and help leaders motivate employees. However, if the Board is anchored in the past, represents legacy organizations, or is loyal to narrow disciplines, a leader must be steadfast in developing ITD strategy. Board members are often eminent leaders with large networks. However, their diversity and power require a subtle hand. They can be aloof, moderately engaged, or deeply involved depending on their defined responsibilities, individual interest, and how well the leader engages them. For example, leaders of Ashoka Trust for Research in Ecology & the Environment (https://www.atree.org/) have been deeply involved with board members as advisors, sounding boards, and fundraisers. Consequently, the organization has built a healthy endowment supporting core staff and functions. This endowment, partly gifted by the board, has allowed the institution to attract reputed faculty, take risks, and be innovative.

EXTRACT 8.3

Pfirman, S. and Martin, P.J.S. (2017) 'Facilitating Interdisciplinary Scholars', in R. Frodeman, J.T. Klein and R.C. Dos Santos Pacheco (eds) *The Oxford Handbook of Interdisciplinarity* (2nd edn), Oxford: Oxford University Press, Chapter 41.

41.2. Institutional Support for Interdisciplinary Scholars

Institutions are recognizing that departmental structures create barriers for scholars working between departments and are adjusting to the needs of interdisciplinary scholars (Table 41.3).

While most institutions have now made at least modest efforts to include interdisciplinary educational programs through establishment of minor courses of study, many others have established interdisciplinary centers and programs, created interdisciplinary departments, and hired senior interdisciplinary scholars. Some have gone as far as breaking down the disciplinary departmental structure altogether (Collins 2002; Feller 2002).

Institutions may create an overlying interdisciplinary framework to provide administrative structures that expand and nurture interdisciplinary collaboration. One example can be seen at Wesleyan University College of the Environment, where students and faculty are supported and linked in a think-tank design that overlays other university units (Poulos et al. 2012). The greatest stress seems to occur at intermediate levels of investment as institutions and individuals attempt to adjust to the needs of interdisciplinary scholars.

Table 41.3. Spectrum of Institutional Interdisciplinary Commitment, Investment, and Therefore Also Responsibility

Commitment and investment	Modest	Intermediate	Significant
Students and curriculum	Minor. General education elective	Concentration. Special major	Major. General education requirement
Administration	Committee	Center. Program	Interdisciplinary department. Dissolution of departments
Faculty	Affiliated hire in disciplinary department. Adjunct hire	Off-ladder. Joint hire	Tenure-track interdisciplinary appointment
Research scientists	Soft-money support for single or short-term project	Multiyear support	Institution-committed career interdisciplinary research scientist line

(adapted from Pfirman et al. 2011).

Because their needs are novel, the scholars often fall between the cracks of administrative responsibility (Figure 41.2).

Being intentional about supporting interdisciplinary scholars requires thinking through the potential challenges in advance. The individual should not be put in the position of having to create their own process at the same time as they are attempting to navigate it. Creating an awareness of differences between interdisciplinary and disciplinary experiences – as we discuss below – can be helpful, from structuring a new hire, to understanding issues related to productivity, teaching, recognition, and evaluation. Awareness, however, is not enough. Funding and administrative support must also be provided. It is critical that institutions make commitments at the level of provost, vice president for research, or dean to the implementation and advancement – not just to the initiation – of interdisciplinarity (Feller 2002).

41.3. Structuring an Interdisciplinary Hire

The process of creating a new interdisciplinary position and the negotiation of the hire often determines the administrative framework of a position, and it is this framework that needs special attention for interdisciplinary scholars. Decisions about new interdisciplinary positions require more extensive cross-institutional preparation than traditional disciplinary hires. At the start of position creation, roles and expectations must be clarified and agreed on by all the departments and academic administrators involved, ideally including representatives of promotion and tenure committees, and those responsible for allocating facilities and resources (Pfirman et al. 2011).

While joint appointments (department-department or department-center) appear to make sense for interdisciplinary scholars, such appointments often lead to difficulties. One is the expectation for service, an expectation that is often double for the joint appointment, serving the needs of two entities -or being penalized for appearing not to serve – for example when teaching is bought out by a research center. Joint appointments may be held to the tenure standards of both departments, which may be at odds (e.g., publications in journals versus books, sole versus multiple authorship). Because responsibility for joint hires is divided, the early career scholar may not get the guidance that they would within a disciplinary department or even through a professional association (Table 41.1). The annual meeting of a discipline's professional association is the place to give presentations, test ideas, and meet the leaders in the field. Interdisciplinary scholars often contribute at the fringe of disciplinary meetings, or risk limited mainstream visibility when they participate mainly in smaller workshops closer to their field of endeavor.

When interdisciplinary faculty are joint hires, it becomes imperative that each department manages their expectations, so that the time and activity demands on the joint appointment are reasonable and not doubled. Having a departmental split of 60:40 or 70:30 may be preferable over a 50:50 split to provide immediate clarity about departmental service (Pfirman et al. 2011). For early career faculty, an even better arrangement might be an "affiliated hire," where they are clearly based in one department but have specific research and teaching contributions to another department, program, or center.

For all interdisciplinary hires, but especially for those who hold joint appointments between departments, the scope of the position should be articulated in a memorandum of understanding (MOU) that spells out scholarship expectations, promotion criteria, teaching responsibilities, departmental and community service, budget, indirect costs, graduate student/technician support, and space (Table 41.1, Figure 41.2). These overarching expectations can then be shared with potential candidates, and later adjusted as part of the negotiation package for the new hire. Interdisciplinary teaching expectations need particular attention. Coteaching classes with scholars from other departments can result in difficult negotiations with the administration and each department about course load, credit, responsibility, content, and pedagogical approaches.

Many interdisciplinary educational goals would be best served by student-centered pedagogy – taking students out into the field, interacting with stakeholders, getting involved in civic engagement, or conducting student-led research. While these types of programs are often cited by students as transformative educational experiences, they are generally considered by the administration to be optional for faculty where the academic program has traditionally been delivered through in-class lectures and structured laboratories. Faculty who choose to incorporate these aspects in their teaching therefore do so at the expense of time they could spend on research, and may even risk having their teaching considered "soft" or "not rigorous" in comparison with colleagues who use more traditional approaches. An interdisciplinary faculty pedagogy forum, joint with schools or departments of education, can be designed to foster sharing of best practices, as well as an increased awareness of the value of new educational approaches and challenges faced within different disciplines. It can also open up education as an area of common ground, building ties between disciplinary and interdisciplinary academic professionals.

An institutional structure that can work well for interdisciplinary hires is a cluster hire (Sá 2008) to support a general theme or initiative, such as environmental sustainability. The administration, relevant departments, and centers work to create the cluster, setting the stage for broad acceptance of the theme. Departments can compete to be the home department of the new hires, thereby creating greater departmental acceptance of the interdisciplinary scholar.

41.4. Productivity and the Interdisciplinary Scholar

One of the most critical aspects to the success for any scholar is their research productivity: the number of publications is often the factor first reviewed for faculty hires and candidates for tenure. Interdisciplinary scholars face hurdles in being productive beyond those of other researchers for a variety of reasons: the field may be new and the scholarly community not yet established, collaborative research requires high overhead/transaction costs in terms of communication, administration (Tables 41.1 and 41.2; Collins 2002; Sá 2008), and additional training requirements. Moreover, each discipline has its own convention for writing grants and publications, and disciplinary-based reviewers often raise issues and request revisions inappropriate for the scope of the interdisciplinary project or difficult to reconcile because they are at odds.

An interdisciplinary scholar can deal with this situation by building expertise in their particular interdisciplinary area – effectively specializing in that area – and then branching into related research topics and publishing in related journals. Researchers in sociology and linguistics who specialized (had a more limited set of key words associated with their publications) were twice as productive as researchers who pursued a research agenda that changed fields substantially over the course of their career trajectory (Leahey et al. 2008).

Although early career researchers in any field are often admonished not to "spread themselves too thin" this advice might be especially important for interdisciplinary scholars. Research by Porter et al. (2007) indicates that scholars who are highly integrative tend not to specialize. It may be that people with a "synthesizing mind" (Gardner 2007, p. 3) use integration as part of their methodology, just as a lab scientist may address research problems using similar instrumentation throughout their career. Spanner (2001) also found that interdisciplinary researchers – especially those at the junior level – were fluid in that they often deviated from their research agenda as they received input from another field.

Börner (2006) has tracked intersections among the disciplines by mapping knowledge domains – in the process creating a communication tool. Interdisciplinary scholars can use this approach to work through related communities in linked networks, expanding their connections and therefore spreading their professional recognition. Mapped knowledge domains not only connect scholarly communities but also can act as another measure of interdisciplinary productivity (Palmer, this volume).

41.5. Recognition of the Interdisciplinary Scholar

Along with productivity, assessment of research performance relies on community reputation, especially recognition for creativity and achievement. Recognition arises from scholars reading and discussing each other's work. Scholars who have achieved recognition serve as informal field gatekeepers, assessing whether a new idea or product should be included in the domain. It is much easier for gatekeepers to recognize innovation when the advance is a direct extension of their own work or that of known colleagues. The difficulty of doing so is compounded by publication in new, interdisciplinary journals with nascent reputations. Without a process and a community for achieving recognition for creativity, the interdisciplinary scholar is faced with significant hurdles in promotion and tenure as well as in funding.

One way to create an interdisciplinary culture on campus, as well as to raise the profile of specific interdisciplinary scholars, is for interdisciplinary scholars to invite leading researchers to give presentations locally. This allows the local scholar to be the host: they get to know the external speaker better, they have the opportunity to talk about their own research, and issues of common interest become something known and talked about on campus. Such interactions are useful for any early career scholar, but are particularly important for those who are interdisciplinary or are in emerging fields. It is helpful in gaining trust if departmental members get a chance to meet prominent interdisciplinary experts firsthand.

While our focus thus far has been mainly on early career interdisciplinary scholars, senior scholars also experience recognition challenges (Pfirman et al. 2011). Most disciplinary societies have something along the lines of a "lifetime achievement award" that identifies major accomplishments and gives credit for accumulated success. In emerging interdisciplinary areas, the scholarly community structures, and therefore the opportunities for recognition, are not well formed. Also, if the interdisciplinary scholar has not specialized, their contributions may be spread over a number of different communities and therefore may not rise to the level of an award in any one of them. Less likely to be the targets of recruitment from other institutions, interdisciplinary scholars may not get the offers that stars do within the disciplines. It is essential that institutions recognize these fundamental differences, and that they support their interdisciplinary scholars – perhaps through the establishment of institutional awards and medals that recognize their overall impact.

41.6. Interdisciplinary Evaluation and Promotion

Conventional, disciplinary-based procedures and standards to assess the work of interdisciplinary scholars ignore the real asymmetries between disciplinary and interdisciplinary research and teaching. In the 2004 Committee on Facilitating Interdisciplinary Research study, concern about "promotion criteria" was the most frequent issue raised by both individuals and provosts in response to a request to rank the top five impediments to interdisciplinary research at their institutions. Mismatched metrics include the number of publications (as noted above, interdisciplinary, multiauthored work often has a slower production rate), focus on single- or first-authored papers (interdisciplinary publication often involves multiple authors), prioritizing well-known, disciplinary journals (not always an outlet for interdisciplinary scholarship), and citation indexes (interdisciplinary research is often new and must build its own constituency).

While tacit knowledge including unwritten guidelines for tenure within a department are passed along through informal collegial interactions and following the outcomes of individual cases, the interdisciplinary scholar is commonly the test case that establishes the criteria through their own performance. But it is not their responsibility to do so – institutions that hire interdisciplinary scholars should create appropriate procedures and metrics, and then be clear about expectations. A compelling way to address this situation is to change how scholarship is evaluated. Boix Mansilla (2006) noted that interdisciplinary work can be viewed through the lens of "consistency with multiple disciplinary antecedents, balance of disciplinary perspectives in relation to research goals, and effectiveness in advancing knowledge through disciplinary interventions" (p. 18). Lattuca (2001) recommends judging all scholarship simply "on the basis of its contribution to the advancement of knowledge" (p. 266).

Another option is to shift from using only "discovery" as the critical component, to the use of Boyer's (1990) expanded set of criteria: "discovery," "integration," "application," and "teaching." Individuals can be asked to provide information on their contributions in each of these areas in their annual performance reports and then the same categories can be used in tenure review. The University of Southern California, Duke University, the University of Michigan Medical School, along with some small liberal arts colleges and some large US land grant universities do this now. However, a word of caution: one study of applied health researchers found that even when interdisciplinarity is at the core of an institution's mission, the chairs of promotion committees, and to a lesser extent the deans, tend to accord significantly more value to traditional scholarly outputs, ranking the importance of nontraditional research output at or below the level of teaching (Phaneuf et al. 2007). Similarly, van Rijnsoever and Hessels (2011) found "that both in basic and strategic disciplines, disciplinary research collaboration is positively related to academic rank, but interdisciplinary research collaboration is unrelated to academic rank".

Reviews of interdisciplinary scholars and proposals can also be facilitated by providing institutional clarity in terms of overall staffing/budget priorities and helping evaluators understand their mission. Letter writers, reviewers, and evaluation committees can be alerted that the scholar or request for proposals is interdisciplinary, and then be provided with the original position or program description. Other options are to collect input from more areas of expertise, permit proposers to provide input on reviewer selection, and allow for proposer response to initial reviews (Langfeldt 2006).

In the case of a tenure review, the make-up of the review committee itself can be critical; it is frequently helpful to include an external expert in the field of the candidate. A problem that can arise, particularly with new areas of interdisciplinary endeavor, is that the outside expert may not be a senior scholar, and therefore may not carry the same professional capital that the external member typically wields in this situation. In order that the review does not depend on this one scholar, individuals can also provide an annotated curriculum vitae, detailing their specific contributions to coauthored publications and grants, cotaught classes, informal advising, and standing of journals/publications – venues that may not be known to members of the committee (Pfirman et al., 2011).

41.7. Funding for Interdisciplinary Research and Education

Traditionally, funding sources, whether internal or external to the university, have been channeled through disciplines. Therefore, support for interdisciplinary research and education is less stable than that for the disciplines. When interdisciplinary proposal calls are issued, they often have incredible competition, typically resulting in funding rates of less than 10%. Then, within 5 years or so, the funding area is often discontinued or moved to another administrative structure. The National Science Foundation, after a period of attention toward an interdisciplinary area, frequently attempts to migrate support back into the core disciplinary directorates, with the goal of changing the culture in the directorates, as well as allowing for new areas of focused attention at the cross-directorate level. However, because of disciplinary pressure, the emerging interdisciplinary areas may not be continued, especially under conditions of budgetary stress.

As a result, interdisciplinary scholars lack continuity in programs and program managers to go to for support. When responsibility for the program shifts, interdisciplinary researchers must establish new contacts, spending considerable effort in rebuilding professional capital. Committing to long-term support – effectively mainstreaming – interdisciplinary programs is essential. Funding agencies and donors can also support research on the reform of faculty reward systems and invest in research on ways to evaluate and facilitate interdisciplinarity.

Funders and institutions can help support interdisciplinary scholars by recognizing that as they initiate interdisciplinary activities, the individual will move "out on the limb" with their infrastructure lagging behind their needs (Collins 2002, p. 81). They can be provided with release time, co-funding, matching funds, and other support for crafting and implementing complex or major research proposals, as well as new interdisciplinary or cotaught classes. Investing and promoting a small number of high-profile projects likely to have success can help institutions develop models that will then reduce resistance to tackling more risky endeavors.

Another major need in terms of funding is to explicitly support all four approaches to interdisciplinarity: intrapersonal, interpersonal/collaborative, interfield/interdepartmental, and working with external stakeholders. The scholar wishing to develop intrapersonal expertise will need seed funding, sabbatical time, and course release as well as perhaps travel support to learn from other institutions, along the lines of the Andrew W. Mellon Foundation "New Directions" grants. Institutions can also provide proposal preparation support: increasingly, requests for proposals are requiring representation from multiple areas of expertise, which often is translated into large collaborative proposals. While both funding agencies and reviewers tend to assume that partnering with multiple institutions always enhances integrative research, for long-term sustainability of research programs it is sometimes better to invest in growing local expertise (Cummings & Kiesler, 2005).

Support is also required to develop opportunities for collegial contact, both professional and social: time and space is needed for collaboration to occur. Co-funding of research centers is one way many institutions are supporting interdisciplinarity. But funding for informal interactions is also helpful. As noted above, most interdisciplinary research is conducted in ad hoc, rather than formal, research teams (Evaluation Associates 1999).

Similarly, 91% of the interdisciplinary scholars in the 2001 Spanner study rated collegial contact as being very important for their work. In addition to serendipitous connections, institutions can build trust through shared experiences such as social occasions and field trips, as well as through the more usual academic paths such as seminar series and workshops. Managing teams is difficult, but managing ad hoc interdisciplinary teams is even more challenging, due to issues conflated with interdisciplinarity (Table 41.2). Explicitly training interdisciplinary scholars in team management could lessen stress and increase effectiveness.

Interdepartmental and interinstitutional initiatives face major hurdles in negotiating terms of budgets, indirect cost recovery, and space. In fact, the Committee on Facilitating Interdisciplinary Research (2004) found that, after promotion criteria, these are the most critical issues faced by interdisciplinary scholars (Figure 41.2). Having a particular person within the institution's administrative structure whose job it is to sort out these issues greatly reduces the transaction costs of initiating new projects.

Commentary

Maureen Burgess and Doireann Wallace

It is widely acknowledged that despite growing interest in inter- and transdisciplinary research, particularly to address major societal challenges, supports are often lacking at institutional level, and inter- and transdisciplinary researchers can be at a disadvantage in pursuing academic careers. Collectively, the three extracts included in this chapter provide insight into how higher education institutions (HEIs) can build support structures and capacity for inter- and transdisciplinary research. Gabriele Bammer et al (Extract 8.1) examine the expertise in integration and implementation required to deliver inter- and transdisciplinary projects and programmes and propose ways to increase and share this knowledge base. Christopher Boone et al (Extract 8.2) outline the kind of leadership needed to establish and drive successful inter- and transdisciplinary organisations. Stephanie Pfirman and Paula Martin (Extract 8.3) address the distinctive challenges researchers face in pursuing inter- and transdisciplinary careers, and make recommendations for facilitating these career pathways at institutional level. From our work as research managers and administrators (RMAs) at Trinity College Dublin, where we have experience supporting researchers in understanding funder requirements, building consortia, preparing proposals and managing funded collaborative projects, we highlight two key points from the texts: the need to understand, map and connect inter- and transdisciplinary expertise within and beyond HEIs, and the importance of long-term vision and leadership to build a culture supportive of inter- and transdisciplinary research.

Bammer et al (Extract 8.1) explore where to find the kinds of research integration and implementation expertise needed to undertake inter- and transdisciplinary research. They note three main sources of such expertise: (1) specific methodologies researchers use, (2) case-based experience acquired through inter- and transdisciplinary practice (frequently tacit) and (3) studies of elements of integration and implementation. They propose developing a global knowledge bank that would coherently index integration expertise for easy access, to help overcome fragmentation and create a 'virtuous cycle', with good practice informing the development of expertise, and successes resulting from this greater expertise in turn increasing interest in capacity building, support for the evaluation of expertise and support for further developing the knowledge bank.

We see a wealth of valuable integration expertise in our own institution, yet it is fragmented. This tacit expertise is often gained on the ground by researchers navigating the messy complexities of inter- and transdisciplinary research projects to effectively manage and integrate the contributions of multiple actors from different disciplines and sectors. Navigating those challenges is almost a rite of passage for emerging inter- and transdisciplinary researchers, and we have seen first-hand how these lived experiences of early inter- and transdisciplinary interactions have informed and enhanced subsequent collaborations. Boone et al's observation (Extract 8.2), that inter- and transdisciplinary leadership emerges and evolves over time, rings true.

It is also worth noting that many staff working in professional roles in research, such as those in the RMA community, have extensive interactional expertise, which Bammer et al in their full article describe as 'the ability to understand disciplines, professional practice and community experience' other than one's own. In supporting collaborative inter- and transdisciplinary research, RMAs routinely navigate the challenges of working with different disciplinary and increasingly, stakeholder groups to develop plans for programmes and initiatives that align with the expectations of funders and policymakers (who also have their own priorities and languages). This constitutes a further significant reserve of tacit knowledge within the institution and should be tapped.

Institutions need to recognise, nurture and facilitate the development of integration expertise, acknowledging the difficult realities of undertaking and supporting inter- and transdisciplinary research (many of which are well described by Pfirman and Martin in Extract 8.3), and finding ways to map and connect the tacit expertise that exists across the institution, among both researchers and professional staff. Bammer et al's proposal to develop a knowledge bank, a shared resource that would make integration expertise more accessible, is important to increase the efficiency of undertaking inter- and transdisciplinary research by highlighting best practice that new inter- and transdisciplinary researchers can build on. From our perspective, some of the most commonly occurring issues that can impact inter- and transdisciplinary collaborations could be helpfully addressed in a knowledge bank through the inclusion of resources that provide guidance on strategically identifying partners and building partnerships, perseverance on the part of leaders and when working with teams, and how inter- and transdisciplinary researchers can best demonstrate the value and success of their methods and projects.

Bammer et al make the important point that indexing and organising resources in the knowledge bank is key (Extract 8.1). Such a resource needs to have multiple entry points to enable use by new entrants (such as early-stage researchers and those from other sectors) as well as more seasoned inter- and transdisciplinary practitioners and scholars. It has the potential to provide entry-level guidance as well as well-structured case studies and more in-depth methods for conducting, evaluating and improving the success of inter- and transdisciplinary projects. Connecting to such a global community of practice would benefit researchers trying to carve out innovative and uncertain career pathways and institutions wishing to support them.

It seems to us that efforts on the part of institutions and their networks to identify and collate integration expertise more formally at local level would be a valuable complement to a global knowledge bank and could strengthen the 'virtuous cycle' Bammer et al refer to through mutually reinforcement at different scales. Institutions would benefit from identifying and sharing sources of tacit knowledge and building an institutional memory around inter- and transdisciplinary theory and practice. This may also contribute to increasing the visibility of inter- and transdisciplinary researchers, who often lack the same processes for achieving recognition as their disciplinary peers (as Pfirman and Martin outline in Extract 8.3). The RMA community can make an important contribution to local efforts to map knowledge across disciplines, helping identify those in the research community with integration expertise as well as bringing interactional expertise from their own experience.

Knowledge sharing within and beyond individual institutions can help build capacity, but there are limits to how much any individual support structure or researcher can achieve alone. For researchers and professional staff alike, such efforts need to be grounded in a longer term strategic vision that supports the creation of inter- and transdisciplinary pathways. Boone et al make the crucial point that long-term vision and leadership are needed to drive culture change (Extract 8.2). Without this, capacity building risks being piecemeal and subject to the interest and goodwill of individuals willing to pursue it, even when short-term priorities seem more pressing. From a researcher perspective, acquiring skills and experience in inter- and transdisciplinary research may be personally satisfying but can lead to frustration when confronted with the reality of the lack of viable career paths (see Chapter 9). Pfirman and Martin outline some of these challenges in Extract 8.3. The fields and journals inter- and transdisciplinary scholars publish in are often less established or their outputs more diverse, making it difficult for disciplinary experts to evaluate their achievements when they compete for funding or jobs. At the same time, inter- and transdisciplinary research requires acquiring skills to integrate different kinds of knowledge to bridge disciplinary and sectoral boundaries. Inter- and transdisciplinary education often requires more time-intensive student-centred pedagogy. Is this additional work rewarded? Is it even visible?

Pfirman and Martin argue that institutional support must begin with an *awareness* of the specific challenges of building an inter- and transdisciplinary career and a *commitment* to put in place the structures needed to enable such careers. They advise that university leadership put considerable thought into inter- and transdisciplinary hiring, ensuring all parties are very clear about responsibilities, promotions criteria and space, to avoid new hires being stretched thin with teaching and research duties across different departments. Evaluation frameworks urgently need to be adjusted to accommodate inter- and transdisciplinary research. Support for inter- and transdisciplinary pedagogy is also needed and can be provided in part, the authors suggest, through establishing faculty-level fora to share best practice. Acknowledging that it takes time to build relationships, they also recommend investments to provide seed funding, time release and physical spaces to enable serendipitous encounters across disciplinary boundaries (a point also made by Boone et al). Circling back to the question of priorities, in our experience the absence of time is one of the most significant barriers to researchers exploring inter- and transdisciplinary research. Increased expectations to develop inter- and transdisciplinary projects add to the burden of work researchers already labour under, and a long-term strategy needs to put structures in place to enable such activity.

Boone et al discuss the leadership qualities and skills needed to establish and drive successful inter- and transdisciplinary research organisations, which applies equally to enabling inter- and transdisciplinary pathways and culture change in HEIs more generally. Inter- and transdisciplinary leaders, they propose, must have a vision that exceeds the status quo, commitment to creating a shared culture and new norms, the ability to develop and nurture strategic partnerships with stakeholders, aligned to long-term strategy, and perseverance. Importantly, they note that these qualities can be present in a leadership team, rather than a single individual. They also recommend investing initially in a small number of strategic projects with a greater chance of success and using these to develop models for wider rollout.

The SHAPE-ID Toolkit includes a case study[1] profiling the Trinity Long Room Hub Arts and Humanities Research Institute, where we have both worked. Founded in 2006, the Hub's journey to build a culture supportive of inter- and transdisciplinary research illustrates the practical value of much of the advice offered in the three texts. The Hub's long-term vision to create a world-class centre of excellence for interdisciplinary arts and humanities research and advocacy has been supported by its ability to secure funding for a flagship building to serve as a home for community development, knowledge sharing and celebrating research across over 20 disciplines. Building on the core strengths and interests of the academic community, this long-term vision has ensured its ability to persevere, experiment, evaluate and continually improve its strategy. Importantly, the case study highlights that culture change takes time, is often incremental and requires the commitment of energetic leadership and an innovative team.

Since its establishment, the Hub has acted as a home for hundreds of early career researchers and driven an ambitious programme of events to bridge disciplinary understandings and reach the public. At the same time, it has continuously explored new ways to enhance cross-disciplinary and cross-sectoral engagement, including hosting public policy fellows, media fellows, artists and writers in residence, and securing strategic funding to build partnerships with industry and civil society, from small-scale workshop grants to European Commission funding for collaborative research and transdisciplinary fellowships. These successes have been enabled by investment in a range of supports, including an internal seed funding scheme with a specific focus on supporting small-scale inter- and transdisciplinary initiatives, a long-established visiting research fellow programme to attract high-profile international scholars who contribute to a vibrant cross-disciplinary community, and specific funded support positions (a research funding officer, research impact officer and communications and marketing officers). Through formal events and the serendipitous informal encounters that the space enables, a virtuous cycle of learning and transformation is created within the community.

The Hub's experience shines a light on an additional feature of creating a culture of inter- and transdisciplinarity. Nurtured in an environment that encourages thinking beyond disciplinary and sectoral boundaries, many Hub researchers have gone on to find rewarding work in industry, policy and civil society organisations. It is worth remembering that most early career researchers will not end up in tenured faculty positions within academia, and supporting inter- and transdisciplinary education and research more broadly can equip graduates with the means and confidence to apply their skills to other challenges, build diverse relationships and pursue other meaningful career paths.

Cultivating integration skills from undergraduate level onwards and committing to creating pathways for inter- and transdisciplinary scholarship, teaching and partnerships should, as the League of European Research Universities (LERU) argue in their position paper (Wernli et al, 2016), become a core business of universities if they are to remain relevant and equip the next generation to tackle the multifaceted and pressing challenges that face global society.

Note

1   See www.shapeidtoolkit.eu/wp-content/uploads/2021/05/Case-Study-Trinity-Long-Room-Hub.pdf

References and further reading

Fam, D., Clarke, E., Freeth, R., Derwort, P., Klanleki, K., Kater-Wettstädt, L., Juárez-Bourke, S., Hilser, S., Peukert, D., Meyer, E. and Horcea-Milcu, A.I. (2020) 'Interdisciplinary and transdisciplinary research and practice: Balancing expectations of the "old" academy with the future model of universities as "problem solvers"', *Higher Education Quarterly*, 74(1): 19–34, DOI:10.1111/HEQU.12225.

Holm, P., Goodsite, M.E., Cloetingh, S., Agnoletti, M., Moldan, B., Lang, D.J., Leemans, R., Oerstrom Moeller, J., Pardo Buendia, M., Pohl, W., Scholz, R.W., Sors, A., Vanheusden, B., Yusoff, K and Zondervan, R. (2013) 'Collaboration between the natural, social and human sciences in Global Change Research', *Environmental Science & Policy*, 28: 25–35, DOI:10.1016/j.envsci.2012.11.010.

i2S (Integration and Implementation) Blog, https://i2insights.org

Lindvig, K. and Hillersdal, L. (2019) 'Strategically unclear? Organising interdisciplinarity in an excellence programme of interdisciplinary research in Denmark', *Minerva*, 57: 23–46, DOI:10.1007/s11024-018-9361-5.

Lowe, P. and Phillipson, J. (2009) 'Barriers to research collaboration across disciplines: Scientific paradigms and institutional practices', *Environment and Planning A: Economy and Space*, 41(5): 1171–84, https://doi.org/10.1068/a4175

Wernli, D., Darbellay, F. and Maes, K. (2016) *Interdisciplinarity and the 21st Century Research-Intensive University*, Leuven: League of European Universities (LERU), www.leru.org/publications/interdisciplinarity-and-the-21st-century-research-intensive-university

9

Developing a Career in Interdisciplinary and Transdisciplinary Research

Chapter overview

Early career researchers in particular can receive very mixed messages about the wisdom of pursuing an interdisciplinary research career. Such careers are often regarded as high risk within a research system that is still primarily structured on discipline-based evaluation and recognition criteria. Success may require a more tactical approach to career development, as discussed in the readings selected for this chapter, which also reflect on why individuals pursue such careers and how institutions need to change and adapt in order to better accommodate – and reward – such endeavours.

A portfolio of skills is essential for success in inter- and transdisciplinary research, especially in an academic context (see also the discussion of 'meta-skills' in Chapter 7). At a minimum, early stage researchers need advice on building skills for interdisciplinary collaboration, extending their mentorship team, bolstering their interdisciplinary CV/résumé for disciplinary review, and preparing for the complications of writing and submitting interdisciplinary grant proposals.

The extract from the article by Dena Fam et al (Extract 9.1) highlights changing expectations of academics in producing alternative research outcomes in collaborative, practice-based research. Through a series of workshops with 20 researchers, these authors identified the tensions behind these outcomes. These reflect the authors' experiences of working in three international sustainability projects, with recommendations for universities seeking to implement inter- and transdisciplinary doctoral and postdoctoral programmes. This piece has broader relevance as it is co-authored by a group of young researchers from different countries and learning cultures.

Recognising that individuals face barriers to their development as effective interdisciplinary researchers, Rachel Kelly et al (Extract 9.2) provide practical advice for early career researchers (ECRs) and their mentors in the form of ten tips. They are presented here to empower present and future generations of interdisciplinary researchers in their endeavour to solve contemporary global challenges.

Helen Bridle et al (Extract 9.3) are a group of early career researchers writing about their experiences of a workshop specifically designed to help shape future interdisciplinary research initiatives. In this extract they reflect on a crucial aspect of interdisciplinary careers: at what stage should researchers and scholars be exposed to this mode of working?

This question of when to 'become interdisciplinary' sits at the core of Catherine Lyall's piece (Extract 9.4), which shifts the spotlight from individual researchers' skills acquisition on to institutions, to explore what practices and mechanisms they can develop to mitigate the constraints, anguishes and fears that an interdisciplinary career may present. Drawing from interviews with interdisciplinary British academics, Lyall describes some of the institutional changes needed to support resilient researchers seeking to craft their own interdisciplinary research trajectory.

EXTRACT 9.1

Fam, D., Clarke, E., Freeth, R., Derwort, P., Klaniecki, K., Kater-Wettstädt, L., Juarez-Bourke, S., Hilser, S., Peukert, D., Meyer, E. and Horcea-Milcu, A.-I. (2020) 'Interdisciplinary and transdisciplinary research and practice: Balancing expectations of the "old" academy with the future model of universities as "problem solvers"', *Higher Education Quarterly*, 74(1): 19–34.

There is an increasing expectation that universities and academic research should collaboratively engage with industry, the economy and/or society (Martin, 2013). This has been widely acknowledged to provide greater possibilities for research impact as well as the need for outcomes that are mutually beneficial to researchers and societal and industry partners (Mitchell et al., 2015). However, the reality of interdisciplinary and transdisciplinary research that is industry and community engaged, is that it is complex, chaotic and 'full of critical moments that disrupt [the] process' (Byrne-Armstrong, Higgs, & Horsfall, 2001: vii).

What this paper has sought to highlight through a process of reflective inquiry is that there are tensions in achieving desired outcomes in interdisciplinary and transdisciplinary research within the academy.

In particular, we have revealed that tensions emerge in the context of a mismatch between the aspirations of the new academy and the legacy of academic systems that still govern and reward disciplinary approaches and research outcomes. In reflecting on our experiences in navigating three recurring tensions, we have identified positive pathways forward and how they might be addressed.

In balancing individual and project team (collaborative) outcomes or what we have referred to as the 'I versus We' dichotomy, we have emphasised the importance of system level change to enable effective leadership which ideally is operationalised both formally and informally through top down and bottom up leadership models to ensure collaborative success is not competing with the personal career development of individual academics and researchers.

In tackling the tension between disciplinary and inter/transdisciplinary outcomes, our research has revealed the importance of ensuring time and resources allocation early on in collaborative projects to ensure shared terminology, problem-framing and collaborative practice is clearly understood by all involved. A recurring theme across the three identified tensions was a need for systems-level change within both university administration and funding bodies to both facilitate and reward the collaborative work, including co-authorship, teaching and team-based collaboration but also in regard to the development of structures to bridge disciplinary silos and facilitate collaboration across disciplinary departments.

In regard to the learning versus research dichotomy, the identification of a need for system-level change suggested a greater need for support for doctoral students/researchers as well as more focus on developing practice-oriented learning opportunities and curricula which might focus on specific interdisciplinary and transdisciplinary competencies required within sustainability sciences (Fam et al., 2017; Meyer et al., 2016). Another overarching recommendation to respond to these tensions is the need to provide opportunities to learn from others, either from experienced researchers or relevant communities of practice working on inter- and transdisciplinary contexts, with specific suggestions of university-wide colloquia to bring together people across faculties to facilitate an open learning culture.

And finally, one of the key characteristics of complex problems, commonly tackled in interdisciplinary and transdisciplinary projects that cannot be overlooked is persistent uncertainty (Head & Alford, 2015), which presents both a challenge and an opportunity. Particularly as we navigate a path towards future models of research and academia, the creative opportunities that these tensions afford must be considered in programmes and the processes of reflection and reflexivity embraced as necessary skills for researchers undertaking such research. What the outcome spaces framework has offered in this research is the opportunity to reflect on the range of tensions that challenge and/or impede academics in identifying and achieving desired outcomes in complex inter- and transdisciplinary projects. In addition, the authors have begun to identify how personal, structural and systemic challenges might be overcome.

EXTRACT 9.2

Kelly, R., Mackay, M., Nash, K.L., Cvitanovic, C., Allison, E.H., Armitage, D., Bonn, A., Cooke, S.J., Frusher, S., Fulton, E.A., Halpern, B.S., Lopes, P.F.M., Milner-Gulland, E.J., Peck, M.A., Pecl, G.T., Stephenson, R.L. and Werner, F. (2019) 'Ten tips for developing interdisciplinary socio-ecological researchers', *Socio-Ecological Practice Research*, 1(3): 149–61.

3. Tips for interdisciplinary researchers

The 10 tips outlined below are presented in a generic manner to increase their applicability and utility across disciplines, geographies, career stages and contexts. However, as outlined earlier, not all tips will be useful across all contexts and some may be more relevant and useful to students and early career researchers, while others to more senior, mentoring or institutional levels. Specifically, tips 1–8 are aimed at the individual researcher and the final tips, 9–10, are focused more on the interdisciplinary team or research strategy level. The tips are presented in an order that logically accompanies the process of becoming an ID researcher, as opposed to the order of the frequency in which they were discussed (i.e. number of experts). Further, these tips are not mutually exclusive, but rather emphasise the key themes of our analysis, which in many cases reinforce one another. We have chosen not to identify specific expert comments and instead use identity codes (e.g. E1 for expert 1, etc.) in the quotations below.

3.1. Tips pertaining to knowledge
3.1.1. Tip 1: develop an area of expertise–work on your core
Interdisciplinarity means bridging between disciplines, but a core grounding is required to bring an expert perspective to the interdisciplinary table comfortably, confidently and most importantly, competently.

> "As oddly contrasting as the terms might be, I think to be interdisciplinary, you also have to be a specialist at something and find that balance." (E3)

This 'core' knowledge can be either a discipline (e.g. marine zoology), a place (e.g. having a deep experiential understanding of the Arctic), a field of study (e.g. fisheries), a method (e.g. modelling) or a process (e.g. knowledge brokering). A core knowledge provides a clear identity and profile. It will shape the researcher's contribution and be a stronghold in interdisciplinary collaborations because developing in-depth knowledge promotes an appreciation of the expertise of others.

Note: Several of the experts believe that 'core' knowledge is contextual and that core grounding can also be IDR. Training as a 'core' ID researcher from the beginning of a research career is challenging, but opportunities for early career researchers to engage in interdisciplinary training are increasing and improving, as we discuss in the next section.

3.1.2. Tip 2: learn new languages–seek to understand and speak across disciplines

IDR requires expression of disciplinary science in ways that are understandable to other disciplines. We are trained to use jargon because it is specific and exact within our fields and/ or disciplines, but in IDR, this jargon will be confusing and excluding. Differences in the use of terms and techniques by various disciplines (e.g. specific statistics, mathematical approaches, qualitative methods, etc.) can also generate confusion and misunderstanding, and the use of discipline-specific terms (e.g. 'significance', 'culture', 'function', 'model', etc.) has great potential to marginalise potential collaborators.

In developing a shared language, the challenge arises in not losing the rigour and nuance of terms, as used in a core discipline. Casual and clumsy use of language can promote, and be a symptom of, clumsy thinking. Just as learning a different (e.g. national) language requires an understanding of grammar, nuance and meaning, learning a different disciplinary language also requires this grammatical and cultural understanding. Superficial ability to function and communicate may come relatively quickly, but fluency takes time, patience and immersion. In the long term, working through language barriers will increase flexibility and adaptability to work across disciplines and can foster the creation of IDR questions and solutions.

> "We need the people who can span the disciplines and kind of speak both languages, sit in that boundary as service translators or be able to think within different frames of reference and thinking as it's really valuable and it is a unique skill set (E5)".

Learning new languages can best be achieved by listening, questioning and more listening. Sharing disciplinary definitions or key introductory texts may be helpful. Communication tools, such as metaphors, analogies and stories, may also be used to present experiences and perspectives to diverse audiences and disciplines. Communication experts and knowledge brokers can facilitate dialogue between disciplines to break down language barriers and support understanding and collaboration.

3.2. Tips pertaining to attitudes

3.2.1. Tip 3: be open-minded–appreciate diversity in perspectives and contributions

IDR necessitates integrating divergent disciplines, and navigating this can be intimidating and even lead to confrontation, particularly for researchers who have been taught within the norms and rules of a single discipline. The key to becoming a successful ID researcher is to remain open-minded; open to learning, open to new ways of doing things and open to collaborations that include new types of disciplinary knowledge and non-academic knowledge. Most importantly, one is to remain humble when engaging with other knowledge-holders. Invite questions, ask them to explain and never be afraid to say 'I don't understand'.

> "People who are humble probably do [IDR] better and more rapidly than people who aren't humble". (E2)

3.2.2. Tip 4: be patient—IDR takes time

Some collaborations work more easily than others, but most frequently, establishing successful interdisciplinary collaborations requires time, and lots of it. IDR is a learning cycle, and transaction time can lengthen when collaborators need to understand different disciplinary cultures, languages and approaches. Time should be allocated for iterative cycles of learning and reflection across all stages of the research process: from the development of the research questions to the solutions that are proposed:

> "Being patient and allowing everybody to learn. I think having patience to allow new kinds of working and being open really, and curious". (E1)

Successful IDR is underpinned by trust among members of the research team, and trust-building takes time because IDR can be an uncomfortable, frustrating space requiring lengthy social bonding processes and effort and patience from researchers. This need for time is one of the main reasons early career stage researchers may find it challenging to engage in IDR, because the academic reward system emphasises the regular and rapid production of publications, particularly in the early stages of a researcher's career.

3.2.3. Tip 5: embrace complexity—it can be stimulating and rewarding

Do not view complexity and differences in approaches as a roadblock. Rather, embrace this complexity and appreciate that every researcher will make a contribution. It would be naïve to underestimate the true complexity of the disciplinary cores you and others are representing, or the complexity of the socio-ecological challenges being tackled. IDR will be difficult, but consider its complexity and associated ambiguity as opportunities rather than barriers. For example, working to understand the 'bigger picture', by combining views and knowledge from several disciplines, provides a richer perspective:

> "There's been a bunch of times where I'm reading a paper from another discipline or struggling through that terminology and suddenly go, 'Oh wow, I never would have thought of that'. You find common patterns across scales that you wouldn't have appreciated otherwise, you can open up doorways". (E2)

Research questions should include input and insight from all members of the research team to provide a clear and shared focus for all, facilitating the contribution of all participants' critical role.

3.3. Tips pertaining to attitudes and practices

3.3.1. Tip 6: collaborate widely—but check your ego in at the door

Interdisciplinary work is integrated and collaborative: bringing people together to harness and discuss their collective expertise. Egos will impede progress because IDR requires hearing and appreciating the views and knowledge of others. Learning nuances and differences in thought and approaches among disciplines is crucial for co-creating research questions and approaches to answer them. Building teams of engaged collaborators spanning multiple disciplines is a powerful strategy for advancing understanding of important socio-ecological challenges. Collaborating broadly fosters joint learning, a significant reward for undertaking IDR, and interdisciplinary team members share collective interest in working on the challenge at hand and in learning from other researchers:

> "Take the journey together…it's not a relay race. It's not like, 'I did something and now I'm going to give you the baton, and now you run the next hundred metres to hand the baton to someone else'. It really should not be a relay race". (E3)

3.3.2. Tip 7: push your boundaries—get comfortable outside your comfort zone
IDR provides opportunity to question how you understand things and to challenge yourself to comprehend something differently. Make attempts, big and small, to get outside of your comfort zone, and deliberately expose yourself to novel perspectives, opinions and ideas:

> "There are whole bodies of theory there that I'm completely ignorant of, and I know I'm ignorant of them. I think in some ways you've got to be quite brave and bold to be interacting in these spaces where you know you know nothing". (E7)

For example, read and attend seminars outside of your discipline, and discuss and share what you learn among your peers. Broadening your disciplinary perspective through IDR can prove enlightening and promote novel and innovative approaches to tackling complex research challenges.

3.4. Tips pertaining to practices
3.4.1. Tip 8: consider if (and how) you will engage in IDR
Interdisciplinary career paths will not appeal to everyone nor do they need to. Many researchers address critical questions within their disciplinary boundaries. Personal skill sets and outlooks will differ between those who aspire to conduct IDR and those who prefer to remain within a single discipline.

> 'Some people want to keep pursuing their [single discipline] and they're very, very good at it, and that's the best road for them'. (E1)

It is important to identify personal aspirations and skills and to consider whether IDR aligns with these career and personal goals. IDR is challenging and requires patience and perseverance, but can provide complex and compelling solutions to the difficult questions and 'wicked' problems around us.

3.4.2. Tip 9: foster interdisciplinary culture—support researchers at the grassroots level
Institutional leaders and senior researchers should foster open atmospheres and safe spaces where interdisciplinary work can be discussed and developed, i.e. where saying 'I don't understand' is supported. Lab leaders should be ready to challenge researchers to explain

their jargon and engage and invite researchers from across disciplines and perspectives into their lab group meetings and events. Achieving interdisciplinary culture relies on lab groups and researchers having the freedom to think and work across disciplinary boundaries.

Institutional leaders can work to adapt internal cultures within their organisations, to ensure that interdisciplinary work is valued, resource allocation (e.g. of time, meeting space and finances) can be granted, and that the formal internal and external recognition necessary for career progression can be provided. Informal encouragement could involve assisting early career researchers to take opportunities that support interdisciplinary skill development or discussing the advantages (and pitfalls) of IDR openly and frequently.

> "Connect them to as many good collaborative people in the different disciplines as possible, so that they've got that supportive network from the start". (E2)

More formal support could involve developing adequate training programmes, and ensuring that these opportunities are more accessible to students across all disciplines who want to understand and practise IDR.

3.4.3. Tip 10: champion researchers—showcase examples of interdisciplinary success

Great IDR deserves recognition akin to that awarded within disciplines for academic contribution and practical output. Leaders and champions can create opportunities and remove barriers by creating and promoting high-profile awards and developing other mechanisms to recognise excellence in IDR. Praise and recognition are central to improving researchers' track records, while they incur the transaction costs associated with working across disciplines. Rewards may also motivate otherwise reluctant disciplinary researchers to participate in interdisciplinary work.

> "I've seen people succeed in interdisciplinarity with very, very different strategies. I think there isn't one way, I think there are multiple pathways". (E5)

EXTRACT 9.3

Bridle, H., Vrieling, A., Cardillo, M., Araya, Y. and Hinojosa, L. (2013) 'Preparing for an interdisciplinary future: A perspective from early-career researchers', *Futures*, 53: 22–32.

3.4. Career stage of participants

A relevant question is at what stage of their career should scholars and researchers attend interdisciplinary encounters. While other literature has considered interdisciplinary exposure and experiences at different career stages [23,24,43,44], we found that little published work has focused on encounters as a mechanism to foster interdisciplinary research at different career stages.

Encounters allow for the exchange of research outputs, plans, and visions. PhD students benefit from interdisciplinary encounters through training in cross-disciplinary communication, and a broadening of their perspectives beyond the often narrow and highly specialized doctorate. Several authors argue that such broadening should, although most often does not, already start at the undergraduate level [17,45,46]. An additional advantage for PhD students of broadening their perspectives, besides better placing their work in context, is that this can inform decisions about future career directions at a stage where it is comparatively easy to change field. Cultivation encounters may thus provide substantial benefit to PhD students, and can constitute an essential element in the training of future researchers [31].

Post-doctoral early-career researchers face pressure to bring in funding and publish academic papers in order to secure permanent positions. As for PhD students, cultivation encounters offer the opportunity to place their work in a larger perspective and define future collaborations and lines of research. Additionally, the networking opportunities offered by such encounters are a useful way for early-career researchers to build confidence in both their own research and in developing working relationships across disciplinary borders.

However, at the post-doctoral stage, cultivation encounters could be complemented by development encounters focused on generating new output, especially projects that lead to long-standing collaboration between researchers from different disciplines. This would help early-career researchers to secure funding and develop wider networks. Crucible participation, for instance, is restricted to early-career researchers and several participants consider that the experience was key to the success of their research careers and critical in securing future funding (personal communication). Sandpit organizers, on the other hand, encourage participation from researchers at all career stages, but note that the attitudes and approach of early-career researchers make them particularly successful participants [42].

Senior researchers may also benefit from interdisciplinary encounters. One example might be new motivation and interest sparked by the possibility of exploring new fields, or new application areas for their research approaches. Like early-career researchers, senior academics are likely to be interested in funding opportunities to cement relationships and deliver interdisciplinary work. Consequently, development encounters with this type of incentive might work well. Finally, encounters at this level allow for leading academics to discuss the future research agenda, feeding recommendations into funding bodies and policy makers.

Overall, it would appear that cultivation encounters are more suited to early-career stages, with development encounters best for more senior researchers. However, regardless of career stage, a researcher's previous exposure to interdisciplinarity might impact upon the choice of appropriate encounter [41]. Those new to interdisciplinarity, at all levels of academia, would benefit from cultivation of ideas, concepts and skills related to interdisciplinarity, and exposure to other disciplines. This is in agreement with Lyall and Meagher [5], who found that researchers at all career stages benefited from reflecting on the process of undertaking interdisciplinary work. It is also consistent with the selection criteria of the more development-focused encounters (e.g., Crucible and sandpits), where the attitude towards, and experience of, interdisciplinarity is considered critical to successful participation.

Another question regarding career stage and encounters is whether participants should all be of a similar career stage or not. Encounters at a similar career stage allow participants to share their common experiences. For example, PhD students could discuss supervision and their experiences in obtaining supervision from various supervisors with different disciplines. Likewise, early-career researchers may discuss the challenges of obtaining a permanent position, publication requirements, and barriers related to cross-disciplinary collaboration. Nonetheless, mixed encounters could enrich this interaction, allowing participants to learn from more senior attendees, which is particularly valued by early-career researchers [47].

A further advantage of mixed encounters is the establishment of contacts that might help PhD students to obtain their first postdoctoral position, early-career researchers to build networks, and senior researchers to identify qualified and motivated candidates to join their research team. With regard to sandpits, the choice of a mixed encounter is justified by an attempt to maximize diversity to "boost the richness of thinking around a topic" [42].

One disadvantage of mixed-career stage encounters is that differences in the existing skills and interests of researchers at different career stages could present a challenge in designing and delivering an encounter suitable for all attendees. For example, experienced interdisciplinary senior academics might not benefit greatly from a cultivation encounter that is highly focussed on training and skills development and, equally, a PhD student might find a development encounter aimed at generating research proposals unsuitable. This illustrates the importance of clearly communicating the aims of the encounter to potential participants to ensure they select appropriate encounters to attend.

The survey results indicate that 85% of respondents were opposed to the idea of restricting participation to those from one career stage. One explanation for this finding could be that participants at the Stresa meeting commented that a valuable aspect of the summit was the opportunity to interact with senior academics. The survey results show that a major benefit of encounters is the chance to learn from more experienced colleagues, and that this opportunity is particularly valued by early-career researchers.

EXTRACT 9.4

Lyall, C. (2019) 'Towards New Logics of Interdisciplinarity', in *Being an Interdisciplinary Academic: How Institutions Shape University Careers*, London: Palgrave Pivot, Chapter 6.

Moving Forward with Interdisciplinarity

Promotion and Reward Structures

While there is ample evidence in the literature of the deterrents to interdisciplinarity (e.g. Blackmore and Kandiko 2011) and the cultural and organisational changes required to address these (e.g. Holley 2009), "the gap between the rhetoric of endorsement and the reality of practice" endures (Klein et al. 2016). Nowhere is this more evident than in university promotion criteria, which are persistently ranked as the highest impediment to interdisciplinary work (e.g. NAS 2005; Tarrant and Thiele 2017), and the reality remains that

> [f]oundations may give grants to imaginative groups of scholars but departments decide on promotions or course curricula (Gulbenkian Commission 1996, p. 97)

I asked awardholders and research leaders whether their institution offered any specific guidance regarding promotion and progression for those who follow a more interdisciplinary route. Terms such as "opaque", "lack of transparency", "generic" and "based on patronage" were used in conversations about universities' promotion processes. While some did describe a promotions and appraisals process that explicitly valued non-standard academic characteristics (VR1) and a process of "educating" promotion panels, so that they were open to considering an interdisciplinary CV "in the right way" (VR7), others (e.g. Fiona) described a metricised basis to promotions with set targets for numbers of publications or amount of grant income. More hearteningly, some interviewees noted updates and revisions to procedures but felt there was still some way to go, describing, for example, a lack of parity in promotion procedures between and within departments (Belinda, Diana). Interviewees called on universities to not just "talk the talk" (Tristan) but to "put their money where their mouth is" (Helena) with respect to promotion procedures, with

> a system for recognising ... what value the interdisciplinary contribution has towards the recognition of merit as an academic. ... explaining what interdisciplinary might mean in terms of outputs related to a person's ... academic standing. (Vera)

Reflecting back on the suggestion that interdisciplinary research is often "slow research" (see Chap. 5), the point was made that university procedures do not adequately acknowledge that, in the early stages, new interdisciplinary collaborations can take longer to establish:

> Within the kind of promotion criteria and the kind of indicators of success … thinking about research grant success and papers that you've written, the criteria that we see here at [current university] at least, there's no mention of interdisciplinarity in there and so if you're expecting X number of papers or number of research proposals, then … that has to be in there somewhere. (Owen)

While there is excellent material on this subject originating from the US (notably Klein 2010; Klein et al. 2016) there is no equivalent of the formal tenure process in UK universities around which much of the discussion of interdisciplinary careers in the US hinges and consequently much less career guidance available for UK academics and their institutional leaders. The University of Edinburgh has been pioneering this area. In 2015, we developed guidance on the consideration of interdisciplinary careers as part of the university's promotion documentation, updating this in 2017 to incorporate additional guidance for assessment of team researchers who may or may not be working in interdisciplinary teams (so-called team science). This publicly accessible document aims to assist both colleagues preparing a case for promotion and those evaluating such cases as members of promotion panels. It outlines ways in which levels of quality can be assessed appropriately and fairly and highlights some of the characteristics and acknowledged difficulties of assessing individual contributions to research activity when individuals are working across traditional discipline boundaries.

Role Models, Mentors and Champions

If we are committed to the advancement of interdisciplinary careers how can we best achieve this? Reward and recognition systems are pivotal but this is not the whole story. A system-wide approach that spans the whole career life cycle and recognises that it is not as simple as "fixing" the promotion rules (Klein and Falk-Krzesinski 2017) is called for. For example, Graybill et al. (2006) call for advice and support in how to develop professional identities, juggle multiple commitments and present their skill sets to future employers and discuss how these are equally essential to an interdisciplinary post-doctoral researcher as the training they typically receive from PIs in the disciplinary requirements of their research projects.

The importance of mentors and role models in providing advice and creating the right environment for interdisciplinarity to thrive was highlighted by awardholders (e.g. Carina, Erica, Norman) while others (e.g. Diana, Katya) spoke specifically about the importance of having a mentor who was independent from the supervisory or line management team.

Others were less fortunate: Louisa had not experienced mentoring and the staff annual progress reviews "often don't happen unless someone insists on them" while in Quentin's case he had sought mentoring informally as his university (along with others in the data set) appeared to conflate mentoring unhelpfully with staff annual review.

A generational issue hinders developments here: Fiona asserted that effective mentors truly understand the nuances of interdisciplinarity but felt that this was less common at the higher levels of academic leadership or management where there may be fewer advanced interdisciplinary researchers to serve as role models.

If a university were to establish such a mentoring scheme, there is a risk that it might place too great a reliance on certain experienced individuals. An alternative would be to develop national networks of interdisciplinary mentors through funding bodies or professional organisations such as learned academies. Interdisciplinary scholars may also benefit from multiple mentors across different disciplines or consider "non-traditional" mentors (Fischer et al. 2012).

On this question of role models, the LERU (2016) report debated whether institutions should appoint interdisciplinary "champions" at the level of Vice Rector. From my conversations with a sample of these Vice Rectors, it was evident that they did not welcome this suggestion. Nevertheless, what I did discern in some of these research leader interviews was a lack of nuanced appreciation of what interdisciplinary research might constitute within the arts, humanities and social sciences compared with interdisciplinarity within the natural and medical sciences. As noted before, from conversations with Gina and Julia, it was also evident that their universities did not maximise the potential of capitalising on the interdisciplinary expertise of their own staff. So perhaps rather than the high-level "interdisciplinary champion" mooted by the authors of the LERU report, a more effective strategy would be a network of champions or "super mentors" at different levels of university governance who could advise and coach staff to demonstrate consistent support and commitment to interdisciplinary careers.

One of the mechanisms suggested in interviews with supervisors as part of the evaluation of the original studentship schemes (Meagher and Lyall 2005, p. 36) was "streamlining internal university procedures (e.g. registry), including perhaps the appointment of a senior official of the university with responsibility for interdisciplinarity university-wide". Martin and Pfirman (2017) recommend something similar either in the form of an administrative committee or individual. So perhaps what is also apposite is a senior figure to act rather like an interdisciplinary ombudsman to ensure that university processes provide consistent and fair treatment of staff who work in an interdisciplinary way.

Support for Early Career Researchers

Mentoring is only one aspect of good academic support for early career researchers. Hein et al.'s (2018) survey identified a training gap in, for example, communication and team skills along with a demand for greater networking opportunities with other interdisciplinary scholars to "leverage confidence". This sense of community building was precisely what we recommended to the Research Councils when we evaluated these interdisciplinary studentship schemes (Meagher and Lyall 2005), as it was evident that many of those studentship awardholders were adrift and floundering. Erica knew that publications were required in order to progress but pointed to poor researcher development support and how this varied between departments within the same university:

> I never really know what you can and can't do … when I worked in the [medical department] … that was very different. They didn't care about your personal [development], they cared about the project. They were not interested in you having

a few hours off to go to a talk, you had to ask permission, whereas it's very relaxed in the [social science] department. (Erica)

I asked research leaders if they provided support for early career researchers who follow an interdisciplinary route. Research support in the sense of assistance with grant applications is relatively commonplace but is much less likely to be tailored for interdisciplinary research. Publicity from SR2's university depicts some of their flagship interdisciplinary initiatives as being "instrumental in promoting the academic career of junior scholars" but what she actually described was very distributed support to encourage interdisciplinary initiatives among doctoral and post-doctoral candidates, driven in part by the requirements of a particular funder. Furthermore, this funding initiative did not provide any form of training to support students working in an interdisciplinary way, which we have shown to be in demand with early career researchers who are seeking to develop improved "academic life skills" (Lyall and Meagher 2012).

Echoing previous findings (Meagher and Lyall 2005, 2009), one specific aspect of ECR support that was highlighted by awardholders (e.g. Norman, Gina, Reuben) was the value of dedicated post-doctoral funding for interdisciplinary researchers in order to allow them time to establish themselves as authentic interdisciplinary scholars:

> [I]t's really important for interdisciplinary researchers to be able to have a couple of years to reflect on what they've learned [to] really digest it and to make new links and to continue being interdisciplinary and to build the confidence to remain interdisciplinary because once you get into the institutions as a lecturer you're immediately identified in your title and in the courses that you teach. (Gina)

As part of their commitment to interdisciplinarity, institutions must also recognise that early career interdisciplinarians, and indeed those who are well established, require greater support for wider networking than the average researcher as their cross-discipline interests may take them to a wider range of conferences than simply the annual professional conference of their disciplinary association. Budget allowances could do more to acknowledge that funds for networking are as much the tools of their trade as the laboratory glassware of their monodisciplinary science based peers.

Commentary

Kirsi Cheas

Dooling, Graybill and Shandas (2017: 581) have noted that 'interdisciplinary doctoral students and early career academics have been asked to become "agents of change" and to accommodate "a new academy" in their doctoral training, yet the institutions in which they function have largely not changed'. Likewise, in Extract 9.1 Dena Fam et al note that 'tensions emerge in the context of a mismatch between the aspirations of the new academy and the legacy of academic systems that still govern and reward disciplinary approaches and research outcomes.'

Due to such 'mismatch' between aspirations and reality, there is a delicate boundary between cynicism, realism and reassurance when advising someone who is about to pursue an inter- and transdisciplinary career. With hindsight, there are many things I wish I had known when I entered academia as an enthusiastic student. For instance, I wish someone had warned me that there are silos and power struggles that will seriously affect my efforts to think creatively and prevent collaborations across fields, institutions and regions. At the same time, my struggles have helped me to develop the qualities of perseverance, resistance and tolerance for ambiguity and failure that I find are fundamental for an inter- and transdisciplinary career.

Hence, the argument I develop in this commentary is that while it is very important to prepare inter- and transdisciplinary-oriented students and early career researchers (ECRs) to the challenges ahead, excessive protectionism to prevent them from failing and hence pushing them to pursue a safer disciplinary journey will both kill their creativity and not help make the needed changes in the academic system.

I find that the excerpts included in this chapter are exemplary in maintaining this delicate balance between realism and cynicism, coupled with reassurance; for instance, 'authors have begun to identify how personal, structural and systemic challenges might be overcome' (Extract 9.1). These are indeed the kinds of readings I wish I had been exposed to much earlier on in my career, to be better prepared, but to not give up, either. In what follows, I will discuss some key takeaways from these readings in regards to inter- and transdisciplinary leadership, collaboration and contextual differences, vis-à-vis my experiences as a student and, currently, postdoctoral scholar pursuing an inter- or transdisciplinary career.

Throughout my career, I have struggled to integrate concepts, theories and methods across the fields of Latin American studies, political communication, anthropology, sociology and journalism studies. In my Finnish university, collaboration between these fields was prevented by administrative structures as well as gatekeeping practices maintained by some established scholars in key positions. Feeling alone and miserable, I realised that there needed to be a support system for inter- and transdisciplinary-oriented students in my country. After attending two Association for Interdisciplinary Studies (AIS) conferences, I felt encouraged enough to found Finterdis, the Finnish Interdisciplinary Society, in 2018. The goals of Finterdis are similar to those described by Catherine Lyall in Extract 9.4: creating networking opportunities and community building to leverage confidence, with a special emphasis on the support of students and ECRs pursuing an inter-

or transdisciplinary career. Finterdis is also leading an initiative to create national, regional and global networks of interdisciplinary mentors, likewise inspired by Lyall (Extract 9.4).

Academia is a hierarchical system, and scholars at different career levels often take their roles for granted. Hence, students and ECRs often assume – or are directly told by their superiors – that they cannot be leaders. When founding Finterdis, this happened to me, too; around this time I had barely finished my doctoral degree. Reactions from more advanced and discipline-oriented scholars varied from amusement to admiration and from mere arrogance to kindness and generous support. It was thanks to the latter that I was able to keep going. I thus appreciate Fam et al's emphasis in Extract 9.1 on the 'importance of system-level change to enable effective leadership which ideally is operationalized both formally and informally through top down and bottom up leadership models'.

In Extract 9.2, Rachel Kelly et al cite their interviewee who said: 'People who are humble probably do IDR better than people who aren't humble.' Likewise, Kelly et al note that 'egos will impede progress because IDR requires hearing and appreciating the views and knowledge of others.' I would strongly concur with these observations. Namely, those scholars who, in my experience, have demonstrated the greatest understanding of inter- and transdisciplinary research methods and theory and openness for collaboration have also seemed most genuinely excited to learn from my young Finterdis colleagues and our lived experiences, rather than just offering to advise us from their more advanced positions. I have found that this kind of reciprocal, intergenerational dialogue has not been possible with many heads of departments or directors of disciplines. Many of these powerful academics claim expertise in interdisciplinarity, but their behaviour does not manifest any such capacity – apparent, specifically, in their lack of curiosity toward the views and accomplishments of others. For them, interdisciplinarity seems to be merely a magic word that can be included in solemn speeches and funding applications with the main aim of boosting their egos even more. Thus, my experiences also echo what Lyall (quoting Klein in Extract 9.4) writes about the gap between the rhetoric of endorsement and the reality of practice, with university promotion criteria ranked as the highest impediment for interdisciplinary work.

Universities' failure to recognise the value of interdisciplinarity across different career levels also means that investing time and effort in the maintenance of an interdisciplinary grassroots support community such as Finterdis is likely not going to contribute to the CVs and careers of those involved in it. In other words, the effort to support other interdisciplinarians drained by the university system takes away time and energy from those voluntarily involved in its organisation – time and energy that could be invested in writing articles and proposals that could help guarantee a tenure position in some disciplinary department. As noted by Kelly et al in Extract 9.2, 'the need for time is one of the main reasons early-career stage researchers may find it challenging to engage in IDR, because the academic reward system emphasizes the regular and rapid production of publications, particularly in the early stages of a research career.' At the same time and paradoxically, the existence of bottom-up communities that build on voluntary work without pay also helps universities to abdicate from the responsibilities of providing better care for their inter- and transdisciplinary students and ECRs, because they can assume that this is already being taken care of by grassroots peer support communities. This is why I find it important that grassroots organisations communicate actively with the universities and push them to make improvements.

When I was planning to found Finterdis, some of the cynical comments I received from more advanced colleagues captured this dilemma. They would tell me that if I did manage to create such a community, it would ultimately cause me to burnout, and even before that, it would slow down my own academic progress. This was cynicism with kind intention to protect me from excessive workload. Three years later, I found that Finterdis was, indeed, consuming me and stealing too much valuable time from my research work. Still, if I could go back in time, would I go ahead with Finterdis anyway? Yes. Because I have now witnessed first hand the creation of connections between open-minded people, and the sensation of relief when scholars and students genuinely committed to inter- and transdisciplinary research find each other going through similar obstacles in academia and getting to think of solutions together.

In Extract 9.3, Helen Bridle et al observe that encounters at similar career stages allow participants to share their common experiences. They give an example: 'PhD students could discuss supervision and their experiences in obtaining supervision from various supervisors with different disciplines.' They add that 'mixed encounters could enrich this interaction, allowing participants to learn from more senior attendees'. These kinds of fruitful exchanges are precisely what have occurred in the Finterdis community – ultimately, providing safe spaces where such interactions can pave the way for larger networked communities that are more capable of pushing for a positive change, and sharing the workload so that no one person alone has to become overwhelmed by it. This was also what happened when I announced that I was too exhausted after three years of leading Finterdis: my new colleagues have come to my help. After persistent work, universities and foundations have also increasingly recognised the value of our interdisciplinary community, helping us out in the organisation and planning of activities and projects. In other words, it is good to warn young scholars against taking on anything that may be too much of a burden. But being overly cynical can cause them to drop something that could potentially have made a positive difference.

Finterdis' collaboration with many groups within the new Inter- and Transdisciplinary Alliance[1] has also shown how inter- and transdisciplinary communities across the world benefit from working together. In many ways, this is related to understanding different obstacles to inter- and transdisciplinary careers, and what different regions and institutions can learn from each other to overcome them, rather than trying to reinvent the wheel. One of the SHAPE-ID project's key findings is that there needs to be recognition that the conditions that influence inter-and transdisciplinary research are context-dependent (see the 'Introductory Essay' in this book). For instance, when I have described challenges in my inter- and transdisciplinary career, some US colleagues have asked me why I have not enrolled into an 'interdisciplinary studies' major programme. The answer is simple: my country does not have any, at least similar to those available in the USA. Instead, I have been trying to create such a programme in my country, the founding of Finterdis having been the initial step in this ambitious process.

Likewise, there are country-, university-, faculty- and field-specific challenges. Scholars working in research teams have different career challenges from scholars working alone. Some students and ECRs may be working with colleagues and advisers who do understand and support interdisciplinarity and help create new connections, putting them in a totally different position from those peers who struggle completely alone, with advisers who oppose or do not understand interdisciplinarity at all. As pointed out by Kelly et al (Extract 9.2), language can also be a barrier. Many inter- and transdisciplinary scholars do not have access to key resources in their own language, and lack sufficient knowledge of English or other dominant languages, because their region does not provide that expertise for them. Hence, this kind of feedback ('Why don't you just…?') can be frustrating and hurtful, because it builds on assumptions rather than an ability to recognise the limits of our own knowledge and the privileges we have in our own context but may take for granted. In other words, global exchanges between inter- and transdisciplinary scholars should build on similar humility – a willingness to hear and appreciate the views and knowledge of others (Extract 9.2) – as any inter- and transdisciplinary endeavour.

Seeing positive examples about successful inter- and transdisciplinary careers from around the world can also help us battle cynicism, as we advise the newcomers. According to Dooling et al (2017: 579), the challenges related to jump-starting new inter- and transdisciplinary collaborations may be particularly devastating to early career academics. In their words, 'the reality sets in that some of these divides are best not battled in pre-tenure years'. But how much of our genuine curiosity and ability to think creatively across boundaries even remains at the point of reaching tenure if, until then, we mostly succumb to the legacies of the academic system that rewards disciplinary approaches and research outcomes, or just follow the footsteps of our principal investigators without pursuing new connections of our own? In my view, insisting that our students and mentees should play it safe will only help to reproduce the 'mismatch' between aspirations and reality, giving continuity to structures that are preventing inter- and transdisciplinarity from functioning as we speak. Instead, I believe we need to mentor them, listen to them – and believe in early career colleagues' capacity to make a meaningful change.

Note

1   https://itd-alliance.org

References and further reading

Dooling, S., Graybill, J.K. and Shandas, V. (2017) 'Doctoral Student and Early-Career Academic Perspectives on Interdisciplinarity' in R. Frodeman (ed) *The Oxford Handbook of Interdisciplinarity* (2nd edn), Oxford: Oxford University Press, Chapter 40.

Felt, U., Igelsböck, J., Schikowitz, A. and Völker, T. (2013) 'Growing into what? The (un-)disciplined socialisation of early stage researchers in transdisciplinary research', *High Education*, 65: 511–24. https://doi.org/10.1007/s10734-012-9560-1

Hein, C.J., Ten Hoeve, J.E., Gopalakrishnan, S., Livneh, B., Adams, H.D., Marino, E.K. and Weiler, C.S. (2018) 'Overcoming early career barriers to interdisciplinary climate change research', *Wiley Interdisciplinary Reviews: Climate Change*, 9(5): e530, http://doi.org/10.1002/wcc.530

Klein, J.T. and Falk-Krzesinski, H.J. (2017) 'Interdisciplinary and collaborative work: Framing promotion and tenure practices and policies', *Research Policy*, 46(6): 1055–61, https://doi.org/10.1016/j.respol.2017.03.001

Pfirman, S. and Begg, M. (2012) 'Troubled by interdisciplinarity', *Science*, 6 April, 336 (6077 supplement), www.science.org/content/article/troubled-interdisciplinarity

Rogga, S. and Zscheischler, J. (2021) 'Opportunities, balancing acts, and challenges – doing PhDs in transdisciplinary research projects', *Environmental Science & Policy*, 120: 138–44, https://doi.org/10.1016/j.envsci.2021.03.009

Vienni-Baptista, B., van Goch, M., Lambalgen, R. and Lindvig, K. (in press) *Interdisciplinary Practices in Higher Education: Approaches to Teaching, Researching and Collaborating*, Abingdon: Routledge.

Epilogue

Bianca Vienni-Baptista, Catherine Lyall and Isabel Fletcher

(with contributions from Maureen Burgess, Kirsi Cheas, Nathalie Dupin, Christian Pohl, Sibylle Studer and Doireann Wallace)

Working together on this book made us realise that the writing process had some unusual features that we had not encountered in previous projects. We felt that these reflections should rightly include our contributing authors, and we offered them the opportunity to give us their thoughts and feedback on the process of writing the commentaries. For us, it was important that the commentaries introduced different voices into the book beyond those of the editorial team. Here, we draw on our contributors' observations to share some lessons learned in putting together this anthology of readings, in order to help us all – including our future readers – further reflect on the foundations of interdisciplinarity and transdisciplinarity and together move the field forward.

We all agree that this is a different type of writing from our daily practice in academic settings. We have all of us put a little more of ourselves into the writing than we would normally do. This format gave us the possibility of saying things in 'a new way'. The invitation to comment on selected readings allowed all of us, in our roles as commentators, to start a dialogue with the authors of the extracts. Although asynchronous, these 'conversations' provided us with the opportunity to consider our own scientific and management practices in the light of the selected readings. The writing process exemplified the virtual collaborations with which we have all become more familiar in recent years: we are a network of colleagues, some of whom have worked together closely, and we have each of us met some, but not all, of the other writing team members 'in real life'. But we have not all met together as a writing team for this project, and all the discussions between the editors and the commentary writers were conducted by email. Nevertheless, this writing process could almost be seen as a kind of co-production that included not only the writing team, but also the broader group of authors whose work we have included in this anthology – 'co-production' because it does not require that all those who participate are present in the same room at the same time. We have demonstrated that it is also possible to include the ideas of others through engaging with their publications, as we have done in this book.

In reading the selected extracts as commentary writers we had to think about how the texts we were reading resonated – or not – with our own individual experiences. Interestingly, we had to read the texts differently, with a particular focus in mind, seeking out ideas in the text that we had not seen before. And admitting that our own ideas were often inspired by the work of others.

We found it stimulating to relate our own experiences to the texts, but this exercise was challenging, too. As academic researchers we have a very narrow concept of 'empirical' and do not usually use our personal experiences as a legitimate source of 'data'. When writing about our experiences for the commentaries, we tried to treat them as if these were any other qualitative or quantitative study – always with a grain of scepticism. Commentary writers who are at an earlier stage in their academic careers shared the same feeling. In this case, being invited to examine the seminal texts through the lens of their own thoughts and experiences constituted a 'nice change' from trying to write from the neutral 'outsider's perspective', more typically aligned with a 'scientific view'.

The selected extracts are rich but intricate for a newcomer to the field. Practicalities and good examples were offered in the commentaries to remind the reader that little things make a difference in inter- and transdisciplinary research. By talking about learning and sometimes failure, we ask the reader to be open to learn from our experiences, irrespective of their status and career stage. One of our commentators asked, 'How could we write something that did not just clumsily repeat or summarise those substantial and thoughtful texts?' The answer was to turn our personal limitations into assets by sharing our perspective on how we used such texts at the science–practice interface, and thereby providing practical hints and tips on the texts' relevance for the work that is actually done 'out there' – in other words, sharing the 'wisdom of practice' to use our friend Julie Klein's phrase. It is often difficult to find time for such self-reflection, but Doireann Wallace and Maureen Burgess remind us of the value of stepping back and 'exercis[ing] a different muscle as we tried to bridge theory and practice by looking at the texts'.

We all shared a bit of ourselves in the commentaries – our thoughts and challenges to encourage the reader to feel curious about the texts and topics, and hopefully also the pleasure we found in reflecting on them. We sought to offer readers a sincere and candid account of our own experiences, and have tried to make this book accessible to everyone with a potential interest in this topic with the aim of opening this world up to others rather than simply talking among ourselves. In doing this, we also sought to remind ourselves and others that, amidst so much superficial or instrumental commitment to inter- or transdisciplinarity in solemn speeches, and funding applications, there are a great many colleagues who strive for similar goals and are genuinely committed to the field.

We faced the practical challenges of selecting which readings to include when doubtless other colleagues and scholars working in this field would have chosen others. Sometimes, as Sibylle Studer reminded us in Chapter 3, it can feel as if we are 'chasing a utopia' when we read about inter- and transdisciplinarity. We have tried throughout this book to balance the 'ideal' versus the 'real'. Our commentary writers have helped with that grounding and were a wonderfully responsive and supportive group who stepped up to the challenge and were willing to offer a little piece of themselves. Ethical behaviour should be the bedrock of all meaningful collaborative research, as Nathalie Dupin reminds us in her commentary in Chapter 7: being ethical – and taking the time to treat ourselves and others kindly – helps to create a safe collaborative space for all.

Finally, we have all enjoyed delving into the details of how others frame the topics that structure this book, and finding that some aspects are still made invisible in our daily academic practice. But this allowed us to realise that, notwithstanding all the constraints, the inter- and transdisciplinary research communities have moved forward on questions that are relevant and urgent, such as publishing, funding, evaluation and career planning. This means that many communities sincerely care about inter- and transdisciplinarity, and contain deep knowledge about their practice and theoretical underpinnings. We hope this book inspires the reader to continue along this path and to dig deeper into these topics.

Copyright Permissions

The editors and publishers acknowledge permission from the copyright holders to reproduce the following extracts:

Extract 1.1: Julie Thompson Klein, Chapter 3, 'Typologies of Interdisciplinarity: The Boundary Work of Definition', in Robert Frodeman, Julie Thompson Klein and Roberto Carlos Dos Santos Pacheco (eds) *The Oxford Handbook of Interdisciplinarity* (2nd edn), 2017, pp 24–7, 29–30, © Oxford University Press.

Extract 1.2: Andrew Barry, Georgina Born and Gisa Weszkalnys, 'Logics of interdisciplinarity', *Economy and Society*, 37(1), 2008, pp 26–31, © Taylor & Francis.

Extract 1.3: Philip Lowe, Jeremy Phillipson and Katy Wilkinson, 'Why social scientists should engage with natural scientists', *Contemporary Social Science: Journal of the Academy of Social Sciences*, 8(3), 2013, pp 213–15, © Taylor & Francis.

Extract 1.4: Lisa Lau and Margaret W. Pasquini, 'Meeting grounds: Perceiving and defining interdisciplinarity across the arts, social sciences and sciences', *Interdisciplinary Science Reviews*, 29(1), 2004, pp 54–8, © Taylor & Francis.

Extract 2.1: Catherine Lyall, Ann Bruce, Joyce Tait and Laura Meagher, 'Making the Expedition a Success: Managing Interdisciplinary Projects and Teams', *Interdisciplinary Research Journeys: Practical Strategies for Capturing Creativity*, 2011, pp 74–5; sub-section 'Continuing the Journey'; end-of-chapter questions, © Bloomsbury Publishing.

Extract 2.2: Louise J. Bracken and Elizabeth A. Oughton, '"What do you mean?" The importance of language in developing interdisciplinary research', *Transactions of the Institute of British Geographers*, 31(3), 2006, pp 374–8, © John Wiley & Sons, Inc.

Extract 2.3: Christian Pohl and Gabriela Wülser, 'Methods for Coproduction of Knowledge among Diverse Disciplines and Stakeholders', in Kara Hall, Amanda Vogel and Robert Croyle (eds) *Strategies for Team Science Success*, 2019, pp 115–17, © Springer.

Extract 2.4: James Leach, 'The self of the scientist, material for the artist: Emergent distinctions in an interdisciplinary collaboration', *Social Analysis*, 55(3), 2011, pp 150–2, © Berghahn Books.

Extract 3.1: Felicity Callard and Des Fitzgerald, 'Against Reciprocity: Dynamics of Power in Interdisciplinary Spaces', *Rethinking Interdisciplinarity Across the Social Sciences and Neurosciences*, 2015, pp 101–6, 109–11, © Palgrave Macmillan.

Extract 3.2: Matthias Bergmann, T. Jahn, T. Knobloch, Wolfgang Krohn, Christian Pohl and Engelbert Schramm, 'The Integrative Approach in Transdisciplinary Research', *Methods for Transdisciplinary Research: A Primer for Practice*, 2010, pp 22–3 (two paragraphs: 'It would be a mistake … genuine scientific tasks facing transdisciplinary research'), pp 44–9, © Campus Verlag.

Extract 3.3: Daniel J. Lang, Arnim Wiek, Matthias Bergmann, Michael Stauffacher, Pim Martens, Peter Moll, Mark Swilling and Christopher J. Thomas, 'Transdisciplinary research in sustainability science: Practice, principles, and challenges', *Sustainability Science*, 7, 2012, pp 27–35 excluding table, © Springer.

Extract 3.4: Chris Rust, 'Unstated contributions: How artistic inquiry can inform interdisciplinary research', *International Journal of Design*, 1(3), 2007, pp 71–5 excluding images, © International Journal of Design.

Extract 4.1: Flurina Schneider, Tobias Buser, Rea Keller, Theresa Tribaldos and Stephan Rist, 'Research funding programmes aiming for societal transformations: Ten key stages', *Science and Public Policy*, 46(3), 2019, pp 464–5 (sections 1.3 and 1.4), pp 468–72 (section 3.2), © Oxford University Press.

Extract 4.2: Diana Rhoten, 'Interdisciplinary research: Trend or transition', *Items and Issues*, 5, 2004, pp 6, 9–11, © Social Science Research Council.

Extract 4.3: Julia Stamm, Chapter 19, 'Interdisciplinarity Put to Test: Science Policy Rhetoric vs Scientific Practice – The Case of Integrating the Social Sciences and Humanities in Horizon 2020', in D. Simon, S. Kuhlmann, J. Stamm and W. Canzler (eds) *Handbook on Science and Public Policy*, 2019, pp 380–5, 391–2, © Edward Elgar.

Extract 5.1: Julie Thompson Klein, 'Evaluation of interdisciplinary and transdisciplinary research: A literature review', *American Journal of Preventive Medicine*, 35(2), 2008, pp 121–3 including table, © Elsevier.

Extract 5.2: Katri Huutoniemi and Ismael Rafols, Chapter 35, 'Interdisciplinarity in Research Evaluation', in Robert Frodeman, Julie Thompson Klein and Roberto Carlos Dos Santos Pacheco (eds) *The Oxford Handbook of Interdisciplinarity* (2nd edn), 2017 (hardback), 2019 (paperback), pp 1–4 including Table 35.1, © Oxford University Press.

Extract 5.3: Tom McLeish and Veronica Strang, 'Evaluating interdisciplinary research: The elephant in the peer-reviewers' room', *Palgrave Communications*, 2, 2016, pp 4–5 (section entitled 'Towards solutions'), © Springer.

Extract 5.4: Christian Pohl, Pasqualina Perrig-Chiello, Beat Butz, Gertrude Hirsch Hadorn, Dominique Joye, Roderick Lawrence, Michael Nentwich, Theres Paulsen, Manuela Rossini, Bernhard Truffer, Doris Wastl-Walter, Urs Wiesmann and Jakob Zinsstag, 'Questions to evaluate inter- and transdisciplinary research proposals', td-net, 2008, pp 12–14 (sections 4 and 5), © td-net.

Extract 6.1: Christoph Kueffer, Gertrude Hirsch Hadorn, Gabriele Bammer, Lorrae van Kerkhoff and Christian Pohl, 'Towards a publication culture in transdisciplinary research', *GAIA – Ecological Perspectives for Science and Society*, 16(1), 2007, pp 24–6, © Oekom-Verlag.

Extract 6.2: Christian Pohl, 'From science to policy through transdisciplinary research', *Environmental Science & Policy*, 11(1), 2008, pp 47–51, © Elsevier.

Extract 6.3: Erin Leahey, Christine M. Beckman and Taryn L. Stanko, 'Prominent but less productive: The impact of interdisciplinarity on scientists' research', *Administrative Science Quarterly*, 62(1), 2017, pp 106, 128–31, © SAGE.

Extract 7.1: Julie Thompson Klein, Chapter 2, 'Communication and Collaboration in Interdisciplinary Research', in M. O'Rourke, S. Crowley, S. Eigenbrode and J.D. Wulfhorst (eds) *Enhancing Communication & Collaboration in Interdisciplinary Research*, 2013, pp 19–26, © SAGE.

Extract 7.2: Erin Leahey, 'From sole investigator to team scientist: Trends in the practice and study of research collaboration', *Annual Review of Sociology*, 42(1), 2016, pp 85–9, © Annual Reviews.

Extract 7.3: Myra Strober, 'Difficult Dialogues: Talking Across Cultures', *Interdisciplinary Conversations: Challenging Habits of Thought*, 2011, pp 40–4, © Stanford University Press.

Extract 8.1: Gabriele Bammer, Michael O'Rourke, Deborah O'Connell, Linda Neuhauser, Gerald Midgley, Julie Thompson Klein, Nicola J. Grigg, Howard Gadlin, Ian R. Elsum, Marcel Bursztyn, Elizabeth A. Fulton, Christian Pohl, Michael Smithson, Ulli Vilsmaier, Matthias Bergmann, Jill Jaeger, Femke Merkx, Bianca Vienni-Baptista, Mark A. Burgman, Daniel H. Walker, John Young, Hilary Bradbury, Lynn Crawford, Budi Haryanto, Cha-aim Pachanee, Merritt Polk and George P. Richardson, 'Expertise in research integration and implementation for tackling complex problems: When is it needed, where can it be found and how can it be strengthened?', *Palgrave Communications*, 6(1), 2020, pp 4–5 including Box 1, pp 8–12 without Boxes 4, 5 and 6 but including Figure 1, © Springer.

Extract 8.2: Christopher G. Boone, Steward T.A. Pickett, Gabriele Bammer, Kamal Bawa, Jennifer A. Dunne, Iain J. Gordon, David Hart, Jessica Hellmann, Alison Miller, Mark New, Jean P. Ometto, Ken Taylor, Gabriele Wendorf, Arun Agrawal, Paul Bertsch, Colin Campbell, Paul Dodd, Anthony Janetos and Hein Mallee, 'Preparing interdisciplinary leadership for a sustainable future', *Sustainability Science*, 15(6), 2020, pp 1725–8, © Springer.

Extract 8.3: Stephanie Pfirman and Paula J.S. Martin, Chapter 41, 'Facilitating Interdisciplinary Scholars', in Robert Frodeman, Julie Thompson Klein and Roberto Carlos Dos Santos Pacheco (eds) *The Oxford Handbook of Interdisciplinarity* (2nd edn), 2017, pp 591–8, © Oxford University Press.

Extract 9.1: Dena Fam, Elizabeth Clarke, Rebecca Freeth, Pim Derwort, Kathleen Klaniecki, Lydia Kater-Wettstädt, Sadhbh Juarez-Bourke, Stefan Hilser, Daniela Peukert, Esther Meyer and Andra-Ioana Horcea-Milcu, 'Interdisciplinary and transdisciplinary research and practice: Balancing expectations of the "old" academy with the future model of universities as "problem solvers"', *Higher Education Quarterly*, 74(1), 2020, pp 31–2, © John Wiley & Sons, Inc.

Extract 9.2: Rachel Kelly, Mary Mackay, Kirsty Nash, Christopher Cvitanovic, Edward H. Allison, Derek Armitage, Aletta Bonn, Steven J. Cooke, Stewart Frusher, Elizabeth A. Fulton, Benjamin S. Halpern, Priscila F.M. Lopes, E.J. Milner-Gulland, Myron A. Peck, Gretta T. Pecl, Robert L. Stephenson and Francisco Werner, 'Ten tips for developing interdisciplinary socio-ecological researchers', *Socio-Ecological Practice Research*, 1(3), 2019, pp 152–5, © Springer.

Extract 9.3: Helen Bridle, Anton Vrieling, Monica Cardillo, Yoseph Araya and Leonith Hinojosa, 'Preparing for an interdisciplinary future: A perspective from early-career researchers', *Futures*, 53, 2013, Section 3.4, pp 27–8, © Elsevier.

Extract 9.4: Catherine Lyall, 'Towards New Logics of Interdisciplinarity', *Being an Interdisciplinary Academic: How Institutions Shape University Careers*, 2019, pp 98–103, © Springer.

Index

References to figures and boxes appear in *italic* type; those in **bold** refer to tables.

A

academic 'value' 113
adaptive behavior 157
administrative tasks 72, 165
advisory bodies 192
agenda setting, joint **97**, 101, 102, 114
agonistic-antagonistic mode x, 31–2, 33, 44
Agrawal, A. 177–8, 186–92, 199, 200, 201
Ainsworth, P. 85, 87
alignment of mental models 161
Allison, E.H. 206, 208–12, 220, 221, 222
antidisciplinarity 27–8
appraisals 215, 216
Araya, Y. 206, 212–14, 221
Armitage, D. 206, 208–12, 220, 221, 222
artefact (artifact) x, 84–8
articulation 57–8
arts, humanities and social sciences (AHSS) (general)
 funding calls xix
 roles and functions of 8–10
 scale of research 53
 SHAPE-ID Tool kit xviii, 65
 status of 5, 64, 165
arts, humanities and social sciences (AHSS) collaborations with STEMM
 bridging disciplinary cultures 49, 52–8, 64, 74, 81, 152, 155–6, 169–74, 209
 co-creating research projects 69–73, 75, 84–8, 90, 91, 94
 developing collaborative conditions 60–6

and modes of interdisciplinarity 29–33, 43–5
motivations for interdisciplinarity 33–6, 43–5
power relations 69–73
science policy instrument 109–11, 112, 113, 115, 116
understandings of interdisciplinarity 36–41, 43–5
unstated contributions 69, 84–8, 90, 91
see also publications

B

Baggott, J. 85–6, 87
Bammer, G. 132
 publication culture 137, 138–40
 supporting collaborative research 177–92, 199, 200, 201
Barry, A. 24–5, 29–33, 44
Bawa, K. 177–8, 196–2, 199, 200, 201
Beckman, C. 138, 147–9, 150, 151
behavioral science movement 26
Bergmann, M.
 dimensions and types of integration 68, 69, 73–6, 89–1
 expertise in integration and implementation 177, 178–86, 199, 200, 201
 ideal–typical transdisciplinary research process 76–84, 89–91
Bertsch, P. 177–8, 186–92, 199, 200, 201
boards 192
Boix Mansilla, V. 133, 151
Boon, A. 206, 208–12, 220, 221, 222

www.ingramcontent.com/pod-product-compliance
Lightning Source LLC
Chambersburg PA
CBHW070617030426

42337CB00020B/3826